Transportation Networks and the Optimal Location of Human Activities

TRANSPORT ECONOMICS, MANAGEMENT AND POLICY

General Editor: Kenneth Button, *Professor of Public Policy, School of Public Policy, George Mason University, USA*

Transport is a critical input for economic development and for optimizing social and political interaction. Recent years have seen significant new developments in the way that transport is perceived by private industry and governments, and in the way academics look at it.

The aim of this series is to provide original material and up-to-date synthesis of the state of modern transport analysis. The coverage embraces all conventional modes of transport but also includes contributions from important related fields such as urban and regional planning and telecommunications where they interface with transport. The books draw from many disciplines and some cross disciplinary boundaries. They are concerned with economics, planning, sociology, geography, management science, psychology and public policy. They are intended to help improve the understanding of transport, the policy needs of the most economically advanced countries and the problems of resource-poor developing economies. The authors come from around the world and will represent some of the outstanding young scholars as well as established names.

Titles in the series include:

Transport and Environment
In Search of Sustainable Solutions
Edited by Eran Feitelson and Erik T. Verhoef

Environmental Costs and Liberalization in European Air Transport
A Welfare Economic Analysis
Youdi Schipper

Reforming Transport Pricing in the European Union
A Modelling Approach
Edited by Bruno De Borger and Stef Proost

Travel Behaviour
Spatial Patterns, Congestion and Modelling
Edited by Eliahu Stern, Ilan Salomon and Piet H.L. Bovy

Financing Transportation Networks
David Levinson

Transportation Networks and the Optimal Location of Human Activities
A Numerical Geography Approach
Isabelle Thomas

Transportation Networks and the Optimal Location of Human Activities

A Numerical Geography Approach

Isabelle Thomas

Senior Research Associate, Belgian National Fund for Scientific Research and Professor, Department of Geography, Université Catholique de Louvain, Belgium.

TRANSPORT ECONOMICS, MANAGEMENT AND POLICY

Edward Elgar
Cheltenham, UK • Northampton, MA, USA

Published by
Edward Elgar Publishing Limited
Glensanda House
Montpellier Parade
Cheltenham
Glos GL50 1UA
UK

Edward Elgar Publishing, Inc.
136 West Street
Suite 202
Northampton
Massachusetts 01060
USA

A catalogue record for this book
is available from the British Library

Library of Congress Cataloguing in Publication Data
Thomas, Isabelle
Transportation networks and the optimal location of human activities: a numerical geography approach / Isabelle Thomas.
p.cm. — (Transport economics, management, and policy)
Includes bibliographical references and index.
1. Transportation—Management—Mathematical models. 2. Industrial location—Mathematical models. I. Title. II. Series.

HE147.7.T49 2002
388—dc21 2001051130

ISBN 1 84064 708 6

Printed and bound in Great Britain by MPG Books Ltd, Bodmin, Cornwall

Contents

List of figures *viii*
List of tables *xii*
Acknowledgements *xiii*

1 Introduction 1

1.1 General context 1
1.2 The optimal location problem 5
1.3 The transportation network problem 10
1.4 A geographical perspective 13
1.5 Outline of the book 21

PART I MEASURING THE SHAPE OF THE TRANSPORTATION NETWORK

2 'Measuring' the shape of a network: state of the art 27

2.1 Some definitions 27
2.2 Network indexing in the literature 31

PART II LOCATION-ALLOCATION RESULTS AND THE SHAPE OF THE TRANSPORTATION NETWORK

Introduction to Part II 39

3 Optimal locations and transportation networks: the case of autarky 43

3.1 Introduction 43
3.2 Definition of the inputs 45
3.3 Results of the simulations 49
3.4 Extension to hexagonal lattices 61
3.5 Conclusions 73

4 Optimal locations and transportation networks: the case of a common market 75

4.1 Introduction 75
4.2 Model and experiments 78
4.3 Impact of the transportation network on locations 82
4.4 Concluding remarks and policy implications 92

5 Optimal locations of human activities and the permeability of the border in a common market 101

5.1 Introduction 101
5.2 How far does the permeability of the border influence location-allocation results in a common market? 107
5.3 How far does the existence of an interregional high-speed connection affect location-allocation results in a common market? 119
5.4 Concluding remarks and policy implications 135

PART III LOCATION-ALLOCATION MODELLING AND THE MEASURE OF DISTANCE

Introduction to Part III 139

6 Distance-predicting functions and location-allocation results. 143

6.1 Introduction 143
6.2 Design of the experiment 145
6.3 Computational experience 150
6.4 Conclusions 169

7 Price policies, transportation networks and location-allocation results 171

7.1 Introduction 171
7.2 Methodology 173
7.3 Principles of the computational experiment 181
7.4 Experimental results 182
7.5 Concluding comments 192

8 Negative externalities and location-allocation results 193

8.1 Introduction 193
8.2 Design of the experiment 195
8.3 Experimental Results 201
8.4 Conclusion 211

PART IV LAND-USE PLANNING AND THE SHAPE OF THE TRANSPORTATION NETWORK. TWO REAL-WORLD EXAMPLES

Introduction to Part IV 215

9 Optimal locations of health centres in Niger: rainy season versus dry season accessibility 217

9.1 Introduction 217
9.2 Background and study area 219
9.3 Modelling results 224
9.4 Conclusion 233

10 Optimal location of a recycling centre in Belgium: externalities versus transportation costs 235

10.1 Introduction 235
10.2 The model and its inputs 237
10.3 Modelling results 242
10.4 Concluding remarks 249

11 Conclusion 251

11.1 Main results and policy recommendations 251
11.2 Suggestions for further research 256

References 261
Index 287

List of figures

1.1	Relationship between economic geography and economics	18
2.1	From a complete graph to a structured transportation network	29
3.1	One fundamental cell of network *a*	45
3.2	Schematic representation of the types of networks	47
3.3	Example of circumradial network: h_5	49
3.4	Average distance and type of network	51
3.5	Average distance and number of links	51
3.6	Maximum distance and type of network	52
3.7	Minimum, average and maximum allocation	53
3.8	Optimal locations on regular networks Eucl, a to f	54
3.9	Optimal locations and allocations for the regular network *a*	55
3.10	Optimal locations and allocations for the radial network h_4	56
3.11	Optimal locations and allocations for the radial network h_2	57
3.12	Optimal locations and allocations for the radial network h_7	57
3.13	Distance frequencies to the supply sites for regular networks	58
3.14	Distance frequencies to the supply sites for circumradial networks	59
3.15	Variation of the average distance with the type of network and the number of facilities	60
3.16	Regular networks and examples of radial networks	62
3.17	Average distance and the number of edges	64
3.18	Average distance and the location of the ring road	64
3.19	Optimal location of the 7 supply points and type of network	66
3.20	Optimal locations on network *R6r*	67
3.21	Optimal locations on network *RHr*	68
3.22	Optimal locations and allocations on network $R3_0$	68
3.23	Optimal locations and allocations on network $R6_0$	69
3.24	Minimum, maximum and average workloads and type of network	70

4.1 Map of the mixed network (common market) 81
4.2 Variation of the number of facilities with F 83
4.3 Optimal locations when $F = 80\,000$ 84
4.4 Variation of the number of facilities with F, for 3 networks and 2 types of demand distribution 86
4.5 Variation of the number of facilities with F, for a circumradial network and 3 types of distributions of demand 88
4.6 Variation of the optimal number of facilities with F and the type of network 88
4.7 Optimal locations in a common market for different F 97
4.8 Optimal locations for two autarkies for different F 99
4.9 Optimal locations in a common market (15×15 LHS + 8×8 RHS). $F = 10\,000$ 89
4.10 Optimal locations in a common market (15×15 LHS + 8×8 RHS). $F = 20\,000$ 90
4.11 Optimal locations in a common market (15×15 LHS + 8×8 RHS). $F = 30\,000$ 90
4.12 Optimal locations in a common market (15×15 LHS + 8×8 RHS), demand being equal to $1\ \forall j$. $F = 10\,000$ 91
4.13 Optimal locations in a common market (15×15 LHS + 3×3 RHS), demand being equal to $1\ \forall j$. $F = 10\,000$ 91

5.1 Examples of toy-networks 109
5.2 Variation of the number of facilities, F and with the number of gates 110
5.3 Optimal locations when γ takes the value of 1 and 3 111
5.4 Optimal locations when γ takes the value of 5, 8 and 15 111
5.5 Number of facilities, F and location of the gates 113
5.6 Variation of the optimal locations and 3 ways of locating the gates 114
5.7 Variation of the optimal locations and 2 ways of locating the gates 114
5.8 Variation of the optimal locations and allocations with the length of the gates 118
5.9 Schematic representation of the studied networks 121
5.10 Optimal locations and allocations for two economies bound together by a central link 125
5.11 Variation of the number of facilities with F and τ 126
5.12 Optimal locations and allocations when a TGV link is constructed between the two regional centres 130
5.13 Optimal locations and allocations when a TGV link is constructed and when one network is less dense 134

6.1 Mathematical spheres for some critical values of p 147
6.2 Relative variation with p of the computed l_p-distance and 2 outputs of the *3*-median model 152
6.3 Computed variation of D_{max} with p for the *3*-median model 152
6.4 Optimal locations and allocation changes with p, when $\theta = 0°$ 154
6.5 Variations of the variation coefficient with p when $\theta = 0°$ 154
6.6 Variations of the skewness coefficient with p when $\theta = 0°$ 155
6.7 Variations with p of K when $\theta = 0°$ 155
6.8 Variation with p and θ of the maximum distance 156
6.9 Variation with p and θ of the ratio between maximum and average distances 157
6.10 Variation with p and θ of the minimum and maximum allocation 158
6.11 Optimal location and allocation changes with p 159
6.12 Optimal location and allocation changes with θ 159
6.13 Variation of the variation coefficient with p and θ 160
6.14 Variation of the skewness with p and θ 160
6.15 Variations of the kurtosis with p and θ 160
6.16 Variation with p of the average and maximum distances for an irregular lattice 162
6.17 Variation with p of the average, maximal and expected distances 163
6.18 Optimal locations in an irregular lattice 165
6.19 Variation with α of the average and maximum distances 168

7.1 Comparison of delivered prices under *M*, *U* and *D* 179
7.2 Comparison of delivered prices under *M*, *U* and *m* 180
7.3 Comparison of delivered prices under *U*, *D*, $Z = 2$ and $Z = 3$ 180
7.4 Optimal locations on a radial network, whatever the price policy 184
7.5 Variation of the quantity of demand generated in each demand point by each pricing policy 186
7.6 Difference between the quantity generated by the *M* and *U* policies – *Circ* 186
7.7 Difference between the quantity generated by the *M* and *U* policies – *Eucl* 189
7.8 Difference between the quantity generated by the *M* and *U* policies – *Rect* 190
7.9 Difference between the quantity generated by the *M* and *U* policies – *Rad* 190

8.1 Intensity, extension and the shape of the studied externality 198
8.2 Average distance and the number of facilities 201

8.3 Variation of the average distance with the number of facilities *k*, *Int* and *Ext* 203
8.4 Variation of the expected total cost supported for *Ext* = 200 205
8.5 Optimal location of 5 facilities when *Ext* = 0 and *Int* = 0 206
8.6 Changes with *Int* of the optimal locations (*Ext* = 600) 208
8.7 Changes with *Ext* of the optimal locations (*Int* = 600) 209
8.8 Changes with *Int* of optimal locations (*Int* = 200 to 575) 210
8.9 Changes with *Int* of the optimal locations (*Int* = 600 to 800) 210

9.1 Filingué: 207 demand points, 11 existing facilities 221
9.2 Suggested enhancements of the road system in Filingué 221
9.3 Variation with *k* of the maximum distance in scenarios 1 and 2 225
9.4 Variation with *k* of the average distance in scenarios 3 and 4 227
9.5 Optimal locations in scenarios 3 and 4 when *k* = 20 229
9.6 Variation with *k* of the average distance in scenarios 2 to 5 229
9.7 Optimal locations in scenarios 3 and 5 when *k* = 11 232
9.8 Hierarchies generated under the 4 main scenarios 233

10.1 Optimal location of one recycling centre for increasing *E* 244
10.2 Number of facilities and average and maximal distances 245

List of tables

2.1	Literature about graph/network indexing	32
3.1	Maximum relative difference between solutions	50
3.2	Number of supply points falling out of the 9 'optimal zones' defined in Figure 3.8	54
3.3	Number of supply points falling out of the 7 'optimal zones' designed in Figure 3.19	66
6.1	Maximum differences between the computed D_{ave}, D_{max} and *CV* for some fixed values of θ	161
6.2	Maximum differences between observed values for some fixed values of *p*	162
6.3	Variation with *p* of the selected optimal sites and their allocated demand	164
6.4	Variation with *p*, σ and θ of the 3-median solutions	166
6.5	Variation with *p*, σ and θ of the *9*-median solutions	167
7.1	PLP results with a *Circ* network	183
7.2	Descriptive statistics about the quantity of demand generated at each of the 81 demand nodes	185
7.3	Pearson correlation coefficients between the amount of demand generated in each *i* for each pricing policy	187
7.4	Location results for 5 price policies and 4 spatial structures when $F = 2\ 700\ 000$	188
7.5	Correlation coefficients between the spatial distributions of demand generated by *M* on different networks	189
7.6	Location results for 5 price policies and 4 spatial structures when $F = 1\ 000\ 000$	191
9.1	Road weights used in scenarios 3 and 4	223
9.2	Index of the optimal sites in scenarios 1 to 2	226
9.3	Index of the optimal sites in scenarios 3 and 4	228
9.4	Average distances for selected values of *k* and 3 scenarios	230
9.5	Index of the optimal sites in scenarios 3 and 5	231
10.1	Mains results for $k = 1$	243
10.2	Mains results for $k = 3$	246
10.3	Characteristics of the users of a recycling centre	248

Acknowledgements

The first version of this research was written as a dissertation to attain the title of Agrégé de l'Enseignement Supérieur at the University of Louvain, Louvain-la-Neuve, Belgium (Thomas, 2000), which is equivalent to a second doctoral dissertation. It is here published in a slightly modified version.

Although only my name appears on the title page, there are many contributors to this book that I would like to sincerely thank. Even though I cannot acknowledge all of them, I do wish to thank some of my mentors, advisors, co-authors and colleagues. Their presence can be witnessed at almost every page of the text. I would like to express my sincere thanks and gratitude to the following:

To Dominique Peeters, for his encouragement, guidance and programming assistance at all stages of the dissertation; he is to a great extent responsible for my interest in the subject matter and for the realisation of this book.

To Jacques Thisse, for innumerable things, but above all for sharing with me his wisdom and vast expertise in the fields of spatial economics and location science. Thank you for helping me to see beyond the computer modelling and to recognise the real impact of transportation on regional development and on the location of human activities.

To Hubert Beguin, from whom I sincerely appreciate his guidance over the past 20 years. He has advised me not only academically, but also professionally. His comments and suggestions on earlier drafts helped me achieve a better dissertation.

I also wish to express my thanks to all the members of the dissertation committee: Francis Borceux, Michel Lebreton, Morton O'Kelly, Dominique Peeters, Denise Pumain, Jacques-François Thisse, Eddy Van de Voorde and Henry Zoller for their valuable support and comments, and to Geneviève Vanbellinghen for helping me in the editing procedure.

Acknowledgement is also given to the Belgian National Fund for Scientific Research (FNRS) for providing the financial support for this project.

Finally, but foremost, an undertaking of this kind puts considerable strain on those closely associated with the author. Hence, my fondest thanks for their support, encouragement and patience go to my husband and three children for the many sacrifices they made throughout this project.

1. Introduction

1.1 GENERAL CONTEXT

Let us start with an example that was – some years ago – the starting point of my interest in the topic developed in this book: the planning of the spatial organisation of police patrols within the Belgian provinces (see Thomas, 1993 and 1994, for further details). For administrative reasons, each province was considered separately. From police and census data, it is easy to estimate the demand for the emergency police services both by type and quantity, at every point of the studied area. From transportation studies, travel times, travel distances and other travel attributes between all pairs of points representing the studied area were computed. The problem then consisted in planning the police patrol services, given a limited budget. One item in the budget was the construction of permanent facilities, the other being the total transportation costs for the patrols. The allocation of demand to facilities, and hence the type and size of the facilities, greatly depends on the number and location of the facilities. The problem in this scenario was to define the limits of the service areas around each facility. Travel times were minimised in order to meet the budget, but also in order to be as close as possible to the users; moreover, the service should not be too far away from the users (equity constraint). Several maps were drawn in order to illustrate the results; some clearly showed that service areas elongate along the main transportation axes. This means that transportation axes apparently influence the results of the optimal location models. Hence when planning the spatial structure of facilities, one also has to take into account the possible and mutual effects of the transportation investments. This was the starting point of the analyses reported in this book.

Hence the chapters of this book aim at presenting an account of how far optimal location-allocation results depend on changes in the shape and size of the transportation network, using a geographical perspective. In other words, our general question is: how far does the spatial structure of the transportation system influence the optimal location of human activities?

This book provides insights into some of the many factors creating spatial differences in optimal location and allocation of human activities.

The literature devoted to location-allocation models generally deals with the problem of where to locate one or more facilities in order to achieve some objective function(s). Location-allocation models are concerned with the location of facilities to serve the distribution of clients best. Thus, this set of models both locates facilities and allocates individuals to them. Their interest is based on the commonly known equity-efficiency dilemma. This is obviously a very important family of problems with countless applications; such location problems arise in many design problems, be it the location of facilities, plants, vehicles, people, services or any other function. While the literature on location modelling has produced a myriad of different operational decision models, very few of them have raised the issue of the influence of the transportation network on the operational results and the decision framework (see Section 1.2 for further details).

Optimal location problems often take place within a given transportation system which is often represented by a network with nodes (for example population centres, crossroads or transhipment centres) and links connecting pairs of nodes (for example roads). Let us note that the terms 'graph' and 'network', 'vertex' and 'node', as well as 'edge' and 'link' will here be used interchangeably. In this given context, transportation depends on the characteristics of the nodes (for example demand for transport) and occurs along the links/edges; it is taken to connote the generalised costs of travel encountered by individuals in carrying out their activities or by firms in moving freight. By generalised costs, we mean some combination of monetary outlays, time length and/or efficiency of travel between specific locations. This way of considering transportation explicitly regards travel as generating negative utilities to the trip-maker. These costs are primarily a function of the supply of transportation infrastructure and of the demand for travel; the latter is, in turn, derived from the demand of individuals and firms for spatially distributed activities (such as employment, commercial outlets or residential locations) which generate and attract trips. Generically, these activities are referred to as land-use activities. The particular distribution and level of intensity of land-use activities are the key factors, which delineate the spatial organisation of regional and urban areas.

When transportation costs decline, it is interesting to know what this would mean to urban patterns of spatial organisation, that is, the compact city versus the suburbanisation process and the development of edge cities. What does this new transport and communication systems imply in terms of systems of cities? In terms of regional development? The study of the interrelationships between land-use planning and transportation has already been discussed in urban and regional economics and geography (see Anas,

Arnott and Small, 1998; Banister, 1995; Berechman, 1994; Berechman et al., 1996; Giuliano, 1989; Rietveld, 1994; Wegener, 1998). Today, the urgency of the environmental debate has renewed the interest in this field of research and especially in integrated models of urban land-use and transport (see LU&T project in Martinez, 2000). Results are, however, conflicting: the literature often asserts that changes in the transportation system caused by – for instance – expansion of the road network, will reduce travel time and costs. These effects, in turn, will encourage dispersion of land-use activities and hence, alter existing patterns of travel demand and thus costs (see for example Bonnafous, 1994; Rietveld, 1994). Krugman (1995), however, shows that a reduction in the transportation costs favours spatial concentration. It is not our intention here to open the debate, nor to review this broad field of research. We here just aim at justifying the main objective of this book by situating it in a broader and up to date field of research.

The relationship between transport investment and regional economic growth is in fact a very complex one, not easily summarised by referring to one regional economic theory or another. A reason of the complexity is that transportation infrastructure has both spatial and economic properties. On the one hand, transportation infrastructure has network properties (Rietveld, 1989a) meaning that it has the extraordinary ability to shift market areas and affect communication channels. On the other hand, it is an input into the production of private and public sector goods. Therefore, it affects the socio-economic landscape in ways no single location model, like those of von Thünen, Weber or Hoover, can fully anticipate (Rephann and Isserman, 1994).

The impacts of transportation networks are often discussed at various spatial scales such as the urban, the regional and the national level (see for example Bruinsma and Rietveld, 1993, 1994; Rienstra et al., 1998). At the regional level, cities can be conceived as nodes in a network, the development of which is influenced by the quality of the links (such as roads, railways or canals) and of the transport nodes (for example airports, seaports). Thus, one can study the effect of changes in transport links (for example high-speed rail) on the relative position of cities in the urban hierarchy. Impact of changes in the transport infrastructure on regions can also be analysed when a transport network is connected to a spatial system in terms of zones or regions. The same illustrations can be given within cities, with organisational aspects of the different modes of travel and their link with accessibility and/or rent or with the shape and structure of the city (see Anas, Arnott and Small, 1998; Anderson, Kanaroglou and Miller, 1996; Wegener, 1994). However, at the intra-urban level, congestion matters. Toy-networks are used in this research project, often with no clear reference to the exact scale of the analysis (intra-urban; intra- or inter-regional); the regional scale of analysis is, however, a reasonable background of interpretation throughout most chapters.

Before closing this general discussion about land use and transportation, it is also worth mentioning a last set of researches to which this book can be linked: models of intercitytransport investments were developed by, for example, Kanemoto and Mera (1985) or Sasaki (1992). These models specify a priori the industrial structure of each city and trade pattern, which is a restrictive assumption for analysis of city size distribution. Urban economists have also developed models explaining how the size of each city is determined in the system of cities (for example Henderson, 1987; Kanemoto, 1980; or Krugman 1996). Recently, Abdel-Rahman (1996), introduced intercity transportation cost in a general equilibrium model of cities and analysed the effect of transport cost on each city's industrial structure. His model has, however, not considered spatial factors at the intercity level, such as location of cities, distances or transport costs between them. Therefore, existing models of urban systems are not useful in explaining what type of city is developed at each location, how large it is and how transport network improvements affect its size; hence our intention here is to help answer this question. In a recent paper, Mun (1997) presents a general equilibrium model for a multi-city economy, where cities are linked by a transport network. He conducts numerical simulations in order to investigate how different transport network patterns affect city size distribution and the welfare of household. He limits himself to three cities and three industries and shows that improvement in the transport network (for example, construction of highways) induces concentration of firms and households in a particular city, and that such concentration significantly increases welfare.

We are aware that this research is devoted to only one aspect of this general land-use problem: it analyses the link between the transportation network (characteristics, properties and shape) and the operational results of some optimal location-allocation models, all other sources of variation being held constant. More particularly, analyses are conducted in order to show the 'sensitivity' of the facility location-allocation problem to the underlying network structure and to try to answer the following question: how far does the transportation system (existence of roads, weight associated to the edges) influence the optimal locations of human activities? Simulations on toy-networks will be used in order to try to answer this question.

In fact, the empirical results mentioned at the beginning of this section confirm earlier conclusions already obtained theoretically by Hakimi in 1965 and by Thisse in 1975. This latter author indeed demonstrated that the transportation network induces a mathematical structure on the set of geographical nodes and that:

> On ne peut déduire (...) que la construction de voies de communication principales est, à long terme, une condition suffisante pour attirer les localisations. Elle n'est pas non plus nécessaire. Plus simplement, elle favorise l'implantation d'entreprises dans une région donnée. On peut donc conclure que l'existence d'une infrastructure importante de transport invite plutôt qu'elle ne contraint. (Thisse, 1975, p. 148).

His demonstration was mainly theoretical and limited to very simple modelling conditions. Since then, location and geographical techniques have progressed, enabling us to propose alternative techniques to simulate what would be the link between transportation structure and location-allocation results.

Hence, in this book, two well-known location-allocation models will be applied on toy-networks, that is to say theoretical and hypothetical transportation networks designed on several squared or hexagonal lattices of points. Using theoretical networks enables to control many sources of variation that would be encountered if using real planning case studies. The shape of the network (radial, rectangular), the location of a ring road if any, and the way of measuring the weight associated to the edges, were a priori defined as causes of variations in the location-allocation results; their effects are tested individually.

The remainder of this chapter aims at situating the topic studied here into a broader avenue of locational and geographical issues. Section 1.2 provides a general overview of location theory in order to justify some of the many assumptions made in this study. Section 1.3 contains an introduction to the problem of evaluating the shape of the transportation network. Section 1.4 situates the topic in the recent geographical and economic research trends. Finally, Section 1.5 gives the analytical framework of this book.

1.2 THE OPTIMAL LOCATION PROBLEM

Since the beginning of the human civilisation, mankind has been faced with the important problem of where to locate human activities, and more recently where to locate private and public facilities such as schools, warehouses, police stations, postoffices, firestations, health care facilities, retailing, production plants or recreation centres. On average, location problems can take several aspects: how many facilities should be constructed? Where should they best be sited? What is the size and shape of their service areas? What size should the facilities be? (Peeters, Thisse and Thomas, 1998a; Peeters and Thomas, 2001). For hundreds of years, facility locations have largely been determined qualitatively. Following the advent of operations research in the last 30 years, mathematical models have been developed and used in order to solve these problems quantitatively.

The purpose of the location-allocation models is to find a pattern of facilities providing a service to spatially dispersed clients as well as an allocation of the clients to those facilities in order to optimise one or several economic criteria generally related to the accessibility of the facilities by the clients. Location theory was introduced by Alfred Weber (1909), who considered the problem of locating one facility so that the transportation costs were minimised. Since then, much work has been done (see Schärlig, 1969, pp.15–66); significant progress occurred after 1960 and facility location analysis emerged as a field of research. A comprehensive treatment of network facility location models can be found in, for example, Hansen et al. (1987), Brandeau and Chiu (1989), Mirchandani and Francis (1990), Francis, Mc Ginnis and White (1992), Chhajed, Francis and Lowe (1993), Daskin (1995), Drezner (1995), Jones et al. (1995) or Labbé, Peeters and Thisse (1995). According to Krarup and Pruzan (1990), among the huge number of location models developed in the literature, four have played a dominant role: the p-median problem, the p-centre problem, the uncapacitated facility location problem (UFLP) and the quadratic assignment problem. Two of them will be used in this research: the p-median (also noted k-median) which minimises total transportation costs, and the UFLP which minimises the aggregation of the cost of installation of facilities and the total travel cost. These two models have attracted much attention, and despite the simplicity of their basic underlying assumptions they have provided important and insightful quantitative bases for the investigation of the numerous practical location decision problems. Moreover, the transportation costs are determinant in the objective function of these two models, which is important for the purpose of this contribution.

Unfortunately, the publication of applied results in location-allocation analysis is often limited to confidential reports; a short review of applications can, however, be found in Hodgson, Rosing and Shmulevitz (1993) and in Eiselt (1992). Early attempts in location-allocation modelling were often limited to a single criterion and focused on normative and prescriptive analyses of well-defined location problems. Recent developments capture the complex behaviour patterns of both producers and consumers, but also lead to solving larger problems involving multiple criteria and large numbers of locations (see for example Tewari, 1992). Moreover, in early applications, the models were used as analytical tools in a defined theoretical framework of service planning, for example, to determine an optimal system of service centres, to examine the locational efficiency of a given service system or to determine new locations to be added to an existing system. In recent years, the main focus of the applications is to develop analysis systems that will help decision-makers in their day to day decision-making tasks (see for example Densham and Rushton, 1992a).

Whatever the type of model used, the location-allocation models consist in finding simultaneously three elements (outputs of a location-allocation model): (1) the optimal locations of the facilities to be sited, (2) the allocation of the clients to these facilities (market or service areas), and (3) the size (capacity) of each facility, taking particular account of the economies of scale inherent to the operating costs of such facilities. For some models, the number of facilities to be sited is endogenous, for other it is exogenous. Accessibility, equity and efficiency measures are also provided by the most operational models. These outputs will be of particular interest in this research: their changes with the simulation conditions will provide a measure of 'sensitivity' of the models.

Most inputs of the location-allocation models are bound to the construction of a graph (nodes, edges); the problems related to its construction are typical of any representation of a geographic environment. The nodes are located in space and often represent the location of demand ((x_j, y_j) coordinates). They can also represent other elements such as the crossing of two edges or the locations of the potential supply sites. A weight is associated to each node; this weight represents the quantity of demand that will be allocated to the appropriate supply site. The node often summarises the information related to an area, leading to two well-known geographical problems: data aggregation and the modifiable areal unit problem (MAUP) (for example Openshaw and Taylor, 1979; or Arbia, 1989). The edges between the nodes summarise the information related to the underlying transportation system (shape, length and costs). A weight is associated to each edge; it is often a distance or a cost. The information related to the edges enables computation of the shortest path matrix between the nodes. These edges (design, weights) will be of particular interest in this study. The simulations performed in this book help to increase the operationality of the location-allocation models by allowing better understanding of the relation between the inputs of the model and the operational results.

There are several important issues in the formulation and in the computational aspects of the location-allocation problems. Among these, there are some we would like to highlight before developing our thesis.

Firstly, the continuous formulation suggests that a site may be selected anywhere in the plane; in other words, there is a continuum of potential locations. The discrete formulation approach considers that there are only a finite number of known sites, which are feasible. Most real-world decision problems lead to the formulation of location-allocation problems in a discrete space, because they are difficult to solve in practice and because they are conceptually more realistic. The freedom provided by planar formulations permits very simple computational procedures for solving problems involving only one facility, but complicates the development of operational algorithms for the multifacility case. Moreover, in most land-use planning case-studies, there are only a few potential sites and therefore, decision-makers tend to consider a discrete representation to be a more realistic and accurate portrayal of the problem. Real planning problems are often defined on a transportation network. In many such situations, even if all points of the network are possible locations, only a finite number contain optimal locations. Our approach is hence limited to the discrete location problem.

The measure of distance used in location-allocation problems is also worth underlining: on a plane, distances are often calculated on the basis of the coordinates of the points and a given metric. This contrasts with the computing distances on a network, where an underlying transportation network is predefined. In the latter case, an algorithm is needed in order to compute the shortest path distances on the network. Although very fast polynomial algorithms are available to determine these matrices, the requirement to generate and store all the shortest path distances may need extensive computer storage. Hence, the use of explicit network distances limits the size of the problem formulation amenable to solution. On the other hand, in most realistic cases, the planar distance metrics are only approximations for the real distances. To avoid inaccuracy in distance measure, it is better to use the network distance directly. Thus the modeller has to make a trade-off between two available options: to seek greater accuracy and realism in the distance employed at the cost of increased computational effort and data storage requirement, or to accept reduced accuracy and realism when generating the distances with the benefits of reduced computational costs and data storage requirement. Hence, a discrete approach is considered here and the sizes of the studied spatial environments are often limited.

Secondly, the computational complexity of this kind of model is also worth underlining: when solving location-allocation problems, the number of location patterns to be evaluated is combinatorial in nature. In general, most of the well-studied location problems have been shown to be *NP*-hard; this means that, except for some spatial cases (such as certain location problems on a tree-network), until recently no polynomial algorithm existed to give an exact optimal solution (see Hansen, Peeters and Thisse, 1983, p. 240). Thus proven optimality cannot always be hoped for in cases where no known exact algorithm is computationally capable of handling a given instance, within a reasonable time. Hence, until recently, we were compelled to resort to approximate algorithms; some commonly used approximate approaches include model relaxation, heuristic techniques and data aggregation. In the following chapters, exact solutions are always computable, but the reader will sometimes – if seldom – be faced (1) with cases where no solution was found within a given limit of computing time, or (2) with solutions that refer to local optima.

Thirdly, when dealing with public facilities, it is also worth mentioning the great dilemma to be solved by most optimal location-allocation models: these models often minimise the total transportation costs encountered by the users to the facilities as well as the costs associated with the facility itself. Hence, if accessibility was the only criterion to be optimised, the model would suggest the location of many facilities located close to the users (best equity); if only financial costs were considered, then a very small number of facilities would be located (best efficiency and hence scale economies). Finding a compromise between efficiency and spatial equity is the purpose of most location-allocation models applied to public facilities (see for example Thisse, 1994 or Peeters, Thisse and Thomas, 1998a for further discussion).

A fourth and last interesting question is to be mentioned: how far are the location-allocation results influenced by variations in the inputs of the model? Rushton (1989) underlined the importance of sensitivity analyses when applying location models to real-world problems. In fact, there are several ways in which sensitivity analyses can be pursued. A first approach is to assess the influence of the objective function. This is typically the purpose of multicriteria analysis. Eiselt and Laporte (1995) provide a theoretical discussion of objective selection, while an applied example can be found in Richard, Beguin and Peeters (1990). A second approach consists in evaluating the robustness of the solutions to the model assumptions. For instance, Arnold and Peeters (1995) compare the allocation patterns obtained by a k-median problem, where clients are supposed to patronise the closest operating facility, with the predictions of a spatial interaction model, where site-specific attractions and non-nearest allocations are allowed; the purpose is to evaluate the impact of distance friction on the use of the facilities, thus on their relative sizes. In the same vein, Hodgson and Oppong (1989) investigate the influence of the boundaries on the efficiency of the solutions. A third approach is to investigate the effect of the definition of the inputs such as demand or transportation costs. Labbé, Thisse and Wendell (1991) suggest a framework for studying the influence of weight perturbation in median problems. Directly related to this, is the question of zone definition and aggregation for which there already exist several theoretical, methodological and empirical contributions (see Goodchild, 1979; Casillas, 1987; Current and Schilling, 1987; Daskin et al., 1989; Hodgson and Neuman, 1993; Thomas, 1993; Ruhigira, 1994; Fotheringham, Densham and Curtis, 1995; or Francis et al., 1999). In this book, we investigate a fourth approach. The criteria used for location problems are generally related to the accessibility, hence to geographical distances. One may then reasonably expect that the outcomes of this family of models will depend on the way distances are defined; the specific question of the measure associated to each edge will be developed in Part III of this book. Now, in most practical instances, distances result from measures on a communication network (see Part II). Hence, our objective here is to evaluate how far the morphology of the network influences the solutions of some location-allocation problems. As stated before, the *p*-median model and the UFLP are the only models considered in this approach.

There are already some theoretical results available in the literature; they are summarised in Labbé, Peeters and Thisse (1995). These are typical theorems stating that the optimal solution to a given problem can be confined to a subset of the network. However these theorems are too general. We would like to broaden the scope by considering realistic transportation networks; this can only be achieved by means of simulations on toy-networks because of their combinatorial nature. Moreover, most of the above-mentioned properties only deal with locational aspects and little is known about the allocation aspects linked to the size of the facilities, the shape of the service areas and the differences in accessibility to the facility among clients. In the following chapters, we would like to examine how these location-allocation outputs are related to the morphology of the network.

In this book the problem of the length as well as the position of the edges (links) will be considered and particular attention will be devoted to seeing how far the measure of the values of the weights associated to each edge influences the location-allocation results. We sustain that the topology of the underlying network may have a profound impact on the optimal locations.

1.3 THE TRANSPORTATION NETWORK PROBLEM

A geographical environment is by essence continuous (a plane); it is, however, often represented by a set of nodes and edges limited by a border. The border determines the size and the shape of the studied area. Nodes represent the locations of the human and economic activities; in human geography, these nodes often summarise the information for an area: a node can, for instance, represent a city, a township, a firm or a train stop depending on the scale and the objectives of the analysis. These nodes are characterised by their number, by their geographical location (coordinates) and by the weight attributed to them; this weight can be, for instance, the number of inhabitants, the number of square kilometres, the total workforce, the income, and so on. The edges are the links between the nodes materialised in the physical environment by roads, rivers, canals, and soon. In a network, edges are characterised – among other things – by their number, location, shape or length, the flow attributed to them and their capacity (maximum flow). Their flow and capacity latter can be expressed and measured, for instance, in flows of goods, cars, passengers, capital or information. The design of the links depends on the location of the nodes and on the characteristics of the physical environment.

Networks have been studied from a number of perspectives such as growth pole theory, systems thinking, cybernetics, graph theory, resource dependence approach, transaction cost economics, agency theory, and soon. Kaman (1998, p. 279) classifies the network researchers into four streams: social science, regional science, industrial marketing, and infrastructure and logistics. Despite interaction among these clusters of research, each stream tends to apply its own methodologies and formulate its own concepts. They deal with networks in social space, economic space or geographical space, and few studies include the time element. Our approach will be limited to some physical aspects, relating to transportation and communication infrastructure. These enable us to meet, exchange and transfer information, knowledge and culture; goods and people can be transported and services delivered. Without them, there is no exchange.

Graphs are a common way of representing parts of the geographical environment; transportation networks linking cities together or within cities are common examples of such graphs. Networks are the physical supports of many economic and transportation activities; they also greathy influence the location of most human activities: place of residence, workplace, location of firms, location of publics services and even the location and development of the cities themselves are dependent on accessibility and hence, on the shape of the transportation network.

The design of the network on which human activities are to be located is an issue in itself. There is a broad avenue of operations research models enabling to optimise a network design (see Balakrishnan, Magnanti and Mirchandrani, 1997 or Raghavan and Magnanti, 1997 for a review); these originated about 40 years ago and have been used in several areas such as communication, transportation and distribution. However, if the problem is deceptively easy to describe, it is very difficult to solve. Most of these models are spatial cases of the multicommodity fixed charge network design problem described by Magnanti and Wong (1984). Generally, these models aim at minimising some measure of link construction and flow costs; they rather optimise the transportation flows than the construction of the underlying graph. Well-known network design models include the shortest path problem (Dijkstra, 1959), the minimum spanning tree problem (Kruskal, 1956), the Steiner tree problem (Hakimi, 1971), the network design traffic equilibrium problem (LeBlanc, Morkol and Pierskalla, 1975), the travelling salesman problem (Lawler et al., 1985) or the vehicle routing problem

(Garvin et al., 1957). Let us just cite two examples: in the minimum spanning tree problem we wish to design a minimum cost network that contains a single path from every node to every other node (the network contains all nodes and is connected). In the Steiner tree problem, we seek a minimum cost network that connects a set of designated terminal nodes. The resulting minimum cost network can use nodes, called Steiner nodes, other than the terminal nodes. Because of the importance of the problem in practice and because the underlying optimisation problems are technically alluring, the combinatorial optimisation research community has devoted considerable attention to solving these problems (Raghavan and Magnanti, 1997). The network design problems (NDC) appear both as stand-alone problems and as sub-problems in more complex network design applications. Melkote (1996), for example, presented a model to simultaneously optimise facility locations and the design of the underlying network. Both optimal and heuristic techniques for solving the problems were discussed. His analyses help identifying the trade-offs between facility location and transportation decisions. Motivated by the simple observation that changing the network topology is often more cost-effective than adding facilities to improve service levels, these models have a vast array of applications in regional planning.

Some of the questions addressed by network design models include: how many links should be constructed? Where should they be built (and which nodes should they connect?) How wide should they be, that is, what should be their capacity? Our aim is not here to optimise network design, but to show the relationship between the location of human activities and the shape of the (given) transportation network. Hence, in all simulations developed in the following chapters, the shape of the network is given and decided a priori.

Optimal location models aim at determining the locations and allocations that minimise or maximise an objective function under a set of constraints. As mentioned previously, the main input of these models is a discrete representation of the human and physical environment. This latter is represented by a set of points that symbolise demand and supply sites; transportation between demand and supply occurs along a transportation network represented by a graph (nodes, edges). Hence, it is supposed that the results of any location-allocation model will depend on the characteristics of the transportation network such as the number of roads (edges) and their spatial organisation.

Few attempts have been made in the literature to explore the relationship between optimal location methods (Section 1.2) and network topology (Section 1.3). Although publications in network analysis and in location theory are numerous, Berman, Ingco and Odoni (1992, 1994) were the first to recognise that the relationship between facility location and network topology has been largely ignored. They consider a network on which one or more facilities are already located and then examine how the network can be modified most efficiently in order to improve the location of the facility when the measure of facility performance is the minimax objective. The types of possible network modifications fall into two categories: reductions in the length of existing arcs or additions of some new arcs that are not currently in the network. A set of reduction and addition problems is introduced for which exact or heuristic algorithms are presented. The principal objective of the paper is in defining and formulating the problems and not in testing the efficacy of the proposed solution methodologies. By making certain changes (reductions and additions) to the configuration of the underlying network, they show how the utility of the existing facilities can be improved.

1.4 A GEOGRAPHICAL PERSPECTIVE

To appreciate what contemporary geography can offer to spatial modelling and more particularly to location-allocation models, it is important to sketch the way the spatial modelling has developed over the past 30 years. This enables us (1) to show how far spatial analysis covers a wide range of approaches and that its popularity within the discipline of geography has waxed and waned considerably during the last 30 years, and (2) to situate this research in a broader geographical and scientific perspective.

1.4.1 Geography and Spatial Modelling

In the early 1960s, geographers discovered that many simple statistical methods have important applications in spatial pattern analysis. These techniques – such as for example point pattern analysis, nearest neighbour analysis, correlation or regression – quickly established themselves in the geographer's tool-kit of methods. These 'classical' techniques were later completed by more refined multivariate methods such as the commonly used factor or cluster analyses. They now continue to be widely used in many empirical domains and particularly through methods associated with discrete choice theory or categorical data analysis (see for example Wrigley, 1985; Haining, 1993; Cressie, 1993). This trend is often called 'spatial data analysis'; it is based on the statistical analysis of databases: exploratory analysis, hypothesis testing, modelling techniques, and so on. Bailey and Gatrell (1995) distinguish 'spatial data analysis' from 'spatial analysis' which is more general and conceptual. Nowadays, because of the improvements brought by the GIS technologies (see Section 1.4.2) and by the development of geostatistics (see for example Goovaerts, 1997; Wackernagel, 1998), spatial data analysis is more widely recognised as a field of interest in itself with new methodological tools such as variograms, spatial autocorrelation, kriging, weighted regressions, and so on. This quickly changing environment creates interesting challenges for the future of spatial data analysis as an integral part of the methodological toolbox of the geoinformation scientist (Anselin, 1998) and particularly exploratory spatial data analysis (ESDA) which is a subset of exploratory data analysis (EDA) that focuses on the distinguishing characteristics of geographical data and specifically on spatial autocorrelation and spatial heterogeneity (see Anselin, 1998; Bailey and Gatrell, 1995; Cressie, 1993; Fotheringham, Brundsdon and Charlton, 2000; Haining, 1993).

Another research trend in spatial modelling that also developed in the 1960s is worth mentioning: it was largely associated with mathematical models and their application to land-use and transportation models (see for example Anas, 1987; Batty, 1989; or Bertuglia, Clarke and Wilson, 1994 for a review). A significant conceptual step in this phase is the derivation of spatial interaction models, which use entropy-maximising methods. This gives the gravity model a clearer theoretical foundation and enables mathematical methods to be established within human geography alongside statistical techniques. Much applied work also took place with the development, calibration and application of land-use transportation models in particular with respect to a variety of cities and regions around the world. Although this style of research was later less fashionable, it provided a much clearer understanding of model structures and dynamics and an increased level of confidence to move forward into a new era of applied research.

The 'quantitative revolution' in geography was heavily influenced by positivist philosophies and by the physicalist model of science. The quantitative movement soon coalesced around a set of research objectives and practices deriving from the positivist formulation of the scientific method: observation and measurement, experimental design, hypothesis testing, mathematical modelling and theory development, and the standards of objectivity, replicability and analytic thinking. What held this intellectual culture together, however, was not the methods themselves but the underlying belief that the world studied by human geography is amenable to the same kinds of approaches to description and explanation, if not prediction, as the world of physical science. Moreover, positivist geographers were able to define the whole discipline on the basis of one of the most fundamental notions in physics – space. Geometry, the formal science of space, provided the theoretical foundation on which quantitative geography was built, and cemented the identity of a substantial part of the discipline as the spatial science (see Couclelis, 1998, p. 21ff).

In the 1970s, Birkin et al. (1996) also mentioned a 'quantitative backlash': a number of influences changed the perception of spatial modelling and its contribution to planning. The idea that computer models of land - use and transport might contribute to more rational planning culminated in the 1960s; however, the diffusion of these models faltered soon after the pioneering phase, for a variety of reasons. Critics emanated from liberal as well as radical geographers (see Batty, 1994; Birkin et al., 1996; Lee, 1973; Harris, 1994; Wegener, 1998). One of the most fundamental reasons was probably that these models were linked to the rational planning paradigm dominant in most Western countries at that time. They were perhaps the most ambitious expression of the desire to 'understand' as thoroughly as possible the intricate mechanisms of urban development, and by virtue of this understanding to forecast and control the future of cities or regions. The effects of the criticism directed towards spatial modelling was to cast the field into a dark age, where developments in model-based methods in the early 1980s were only undertaken by a limited group of researchers. However two major scientific events changed this course of history: the emergence of the GIS (see Section 1.4.2) and the renewal of the economic geography perspective (see Section 1.4.3).

1.4.2 Geography, GIS and Geocomputation

In 1980s, geographic information systems techniques emerged, renewing the interest in geographical techniques and spatial modelling in general. If at an early stage, research was mainly limited to the GIS technique in itself, the natural complementarity between GIS and spatial modelling nowadays encourages belief in a great future for spatial modelling. GIS techniques have enabled: (1) a much greater availability and management of data that form the necessary inputs of most spatial models; (2) the development of a range of new methods including micro-simulation, methods based on artificial intelligence concepts such as genetic algorithms and neural networks, cellular automata, fractals, fuzzy sets and competing destination or other interaction hybrids; and (3) a greater interest in spatial analysis from a range of non-academic and non-geographical organisations. The success and rapid spread of GIS both within geography and well beyond have been amply documented. Unfortunately, despite substantial progress, in many cases the use of models to support scientific investigations and decision-making has been hampered by computational complexity and poor performance (see for example Longley et al., 1998; Armstrong, 2000).

Geocomputation is a new and rapidly developing area mainly provided by GIS (see Longley, et al. 1998). Geocomputation is characterised by research-led applications, which emphasise process over form, dynamics over statics and interaction over passive response. Longley (1998, p. 3) asserts that:

> Geocomputational representations are mainly digital 'models', simplifications of geographical reality, at various levels and to varying degrees of abstraction. This process of abstraction by definition discards information, whether the underlying purpose is to model spatial distributions (data models), flows (e.g. spatial interaction models), networks, hierarchies or diffusion. The fundamental technology of GIS has undoubtedly developed and matured in recent years, yet GIS remains a tool, which can be used for a wide range of scientific and general ends. In itself, GIS proved little guidance as to how best spatial problems might be solved. It is obvious that big technical and epistemological questions remain about the ways in which selected facets of geographical reality are discarded or emphasised in the process of model building. Geocomputation provides a framework within which those researching the development and application of GIS technologies are seeking to address and resolve many of these important questions. (...) The spirit of geocomputation is fundamentally about matching technology with environment, process with data model, geometry and configuration with application, analysis with local context and philosophy of science with practice.

Hence, a successful location-allocation application requires a satisfactory representation of the geographical and economic environment of the studied problem: the demands of the clients, the costs related to the implementation and to the operation of the facilities in the various sites, the costs related to the movements of people and commodities, the impact of the nuisance generated by the system, and so on. As this generally goes along with a huge data-collecting step, a major issue in the development of location-allocation software is the design of an interface with a geographical information system, which offers more capability in handling georeferenced data and provides the model with its requested inputs (see for example Miller, 1996; Armstrong, 2000). Such a GIS should also be useful at assessing the outcome of the model. Several sections of this book deal with various aspects of environmental modelling; moreover, by working with toy-networks, location-allocation models and several statistical, graphical and GIS packages, we believe that this book fits well into these new perspectives in geography.

1.4.3 The 'New' Economic Geography

Besides the geocomputational aspects, this book can also be fitted into another new perspective in spatial research: the 'new economic geography', which by essence – despite the numerous discussions – belongs to economics as well as geography.

The relationship between economics and economic geography has long been an asymmetric one: mainstream economic theory has concentrated primarily on determining who produces what and for whom. It traditionally explains differences in production and consumption structures through differences in underlying characteristics. The spatial organisation of the economic activities has often been a secondary consideration. Land use, spatial competition and the formation of cities are topics that were, until recently, seldom covered by economic theory. It is as if production, consumption and exchanges take place in a single, dimensionless space, even if these activities consume land and are unevenly distributed over space. This criticism is not new, as Isard (1956) had already spoken of 'a wonderland without dimension'. Transport economics is one sub-discipline within economics that began to develop about 30 years ago with the growing problems of urban transport and the recognition that land use and transport needed to be considered together. The main 'tools' of the transport economist are taken directly from the tool-kit of standard microeconomic theory (see for example Button, 1993).

Mainstream economic geography, on the other hand, sets out to explain the uneven distribution of economic activities; it focuses on the spatial organisation of production, distribution and consumption activities. The economic geographer wishes to display the spatial distribution of these economic activities and to understand why they are located as they are and what flows they generate. A significant part of this understanding lies in recognising how economic activities are interrelated in particular areas and how the activities are tied to other economic activities at other locations (see for example Wheeler and Muller, 1986, p. 4ff). Taaffe (1974) illustrated the relationship between economic geography and economics (see Figure 1.1). The two fields (economics and economic geography) have indeed separated core interests – economic theory and spatial economic analysis, respectively – but share common research concerns (see also for example Lee and Wills, 1997). The result is that, in this overlapping zone, research in regional economics, transport economics, transport geography and locational analysis draw from the ideas and approaches of both economists and economic geographers; these are the disciplines our approach is concerned with.

In recent years, however, a growing number of economists have become interested in the study of location problems leading to a new research trend named 'new economic geography', better called, geographical economics (see Fujita and Thisse, 1996; Krugman, 1998; Ottaviano and Puga, 1998; Clark, Feldman and Gertler, 2000). Although much criticised by some geographers (see Martin and Sunley, 1996; Martin, 1999), Krugman's work has been very influential in promoting this new research trend by the economists. New economic geography focuses on the spatial agglomeration of economic activities and the long-run convergence or divergence of regional incomes.

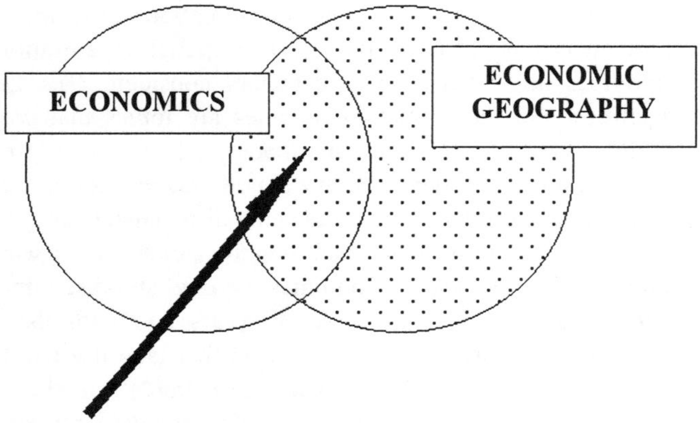

Development economy
Regional economy
Regional science*
International trade economy
Locational analysis*
Theoretical economic geography
Transport economics*
Transport geography*

Note: * refers to the discipline to which this research is directly related

Source: Derived from Taaffe (1974)

Figure 1.1 Relationship between economic geography and economics

According to Krugman, in a world of imperfect competition, international trade is driven as much by increasing returns and external economies as by comparative advantages. Furthermore, these external economies are more likely to be realised at a local or a regional scale than at national or international level. To understand trade, therefore, Krugman argues that it is necessary to understand the processes leading to the local and regional concentration of production. To this end, he draws a range of geographical ideas from Marshallian agglomeration economies, through traditional location theory, to notions of cumulative causation and regional specialisation (Martin and Sunley, 1996). Krugman has written on a wide range of issues that impinge on the regional development question: trade, externalities, the localisation of industry, strategic industrial policy, globalisation, the role of history and 'path dependence', and the implications of economic and monetary integration of regional growth (see Krugman, 1991a, 1991b, 1993, 1995, 1996, 1998). Essentially, this research trend covers three main research programmes concerned respectively with the spatial agglomeration of economic activity, the dynamics of regional growth convergence and a third strand based on the neo-Marshallian district economics. Several leading names are associated with this new trend including Arthur, Fujita, Porter, Thisse and Venables.

It is however fair to say that this new research field is in many respects more rooted in standard economic theory than in the traditional theories or location of geography: some authors even say that it seems that 'economists are discovering geography' (Martin and Sunley, 1996, p. 60). I would personally say that – even if I totally agree with some of Martin's arguments – the economic and geographical approaches turn out to be complementary rather that contradictory, and that the 'new' economic geography helps the 'traditional' geographer to better understand the observed spatial organisation. The new economic geography stems from a failure of classical economic theories to explain observed spatial realities (agglomerations, and so on.), while it gives a theoretical framework to the observation-driven economic geographer.

Several aspects of this new economic research trend appear in this book, which lies somewhere at the border between geography and economics: we mainly analyse the geographical pattern of economic activities as well as the generated flows and also interpret our results in terms of agglomerations (creation, evolution). Agglomerations, scale economies, transportation costs, spatial competition, externalities, and so on. are used in most chapters in order to explain the observed spatial differences in the optimal location-allocation results. Taking one kind of model into account (location-allocation models), we use simulations on toy-networks as well as empirical work in order to describe (classical economic geography) but also to scientifically explain (spatial economy) the observed differences.

Let us however remember that many of the simulations performed in the following chapters rely on the trade-off between scale economies in production and transportation costs, which is central in classical economic geography since it already appears in the pioneering analyses of Christaller (1933) and Lösch (1940), as well as in the 'new economic geography' research program catalysed by Krugman.

In Krugman's setting, there are two regions with no spatial extension and two sectors (agriculture and industry). Industrial workers and firms are mobile while farmers are immobile. Industrial firms sell a differentiated product under monopolistic competition, increasing returns and positive transport costs. Agricultural firms produce a homogeneous good under perfect competition, constant returns and zero transport costs. Finally, consumers (that is, workers and farmers) like to diversify their consumption of the industrial good among many varieties. The main outcome of this strand of literature is as follows: 'high' transport costs lead to the dispersion of the industrial sector because farmers' demands for varieties act as a strong centrifugal force; sufficiently 'low' transport costs yield agglomeration of the industrial sector within a single region because firms may exploit the scale economies associated with a large local market while losing little business in the peripheral region. When workers are not mobile across regions, the existence of inter-industry linkages provides the reason for the concentration of industrial firms within a single region, at least when transport costs are 'not high'. However the concentration of the intermediate and final sectors within one region exacerbates competition for the local labour market, thus boosting higher wages. Hence, sufficiently strong decreases in transport costs eventually lead to the redispersion of industry (Venables, 1996).

Although we here focus on the same trade-off as Krugman, we should make it clear that our research strategy is different. Specifically, we consider a completely different institutional setting in which production facilities are located by the planner, who has all the required information. Our approach is mainly supply-driven. Since the local requirements for the composite good are fixed and given, the planner aims at locating the facilities in order to minimise the sum of production and transportation costs. In other words, we do not deal with market competition, thus making the comparison of the two approaches fairly problematic.

As a possible line of defence, we would like to say that we expect the results obtained by applying location-allocation techniques such as the SPLP to be suggestive of more general tendencies, since the trade-off between increasing returns and transport costs is central in most models of economic geography. Furthermore, in our setting, each region has a spatial extension over which consumers are dispersed. This allows us to study the intra-regional distribution of production when transportation costs vary, a task that has been so far out of reach given models belonging to the 'new' economic geography. At best, our modelling strategy should be viewed as a first step toward a more ambitious programme accounting explicitly for the description of market interaction and the determination of the equilibrium prices on which firms and consumers base their locational decisions. Let us here mention some recent advances proposed by Harker (1993), Nagurney (1993), Nagurney and Zhang (1996) or Labys and Yang (1997) in parallel computation of spatial and temporal price equilibrium models. They make complementarity and variational inequality formulations more attractive candidates for multi-commodity or multi-period inventory adjustment specifications and open up a broad avenue for further research. However their combination with location-allocation techniques is, in our opinion, a very difficult task because it leads to quite non-convex problems that are far beyond the objective of this book.

In our approach, demand for transport is also implicit; it is derived by the location choices, but it is not modelled (see Ben Akiva and Lerman, 1985 for adequate modelling procedures). We have limited ourselves to the planning approach, which should further be integrated in order to include equilibrium processes where location, consumption and production are endogenous and depend upon the market processes.

1.5 OUTLINE OF THE BOOK

As mentioned in the preceding sections, facility location analysis has been extensively used for siting public and private facilities; however, in the literature, few attempts have been made in order to explore systematically the relationship between facility location and network morphology. This will be the main focus of the present contribution.

Hence, the rest of this book is divided into four parts of unequal length. Part I is limited to one chapter (Chapter 2) and is devoted to the definition, the description and classification of networks. Given the final objective of this book, it would be extremely useful if we could 'synthesise' unambiguously the morphology of a given transportation system by one or several quantifiable indices. These indices should be conceptually descriptive, easily measurable and suitable at a variety of scales. Hence, we will first try to attribute to each graph or network one or several indices in order to best 'describe' the spatial organisation of the nodes and links (the shape of the network) and to cluster the networks in exclusive categories. Despite large hopes, Part II concludes that even with the new technologies such as neural networks or fractals, it is still very hard – if not impossible – to characterise a network by one or even several indices. Hence, subsequent developments will here be oriented towards simulations on toy-networks, without any reference to the quantitative 'description' of the network.

Part II strictly deals with the link between the shape of the transportation network and the location-allocation results; it is perhaps the most innovative point of this book. Our objective is here to evaluate how far the morphology of the network influences the solutions of location-allocation problems. We limit ourselves to two optimal location models: the k-median and the SPLP (see Section 1.2). Hence, Chapter 3 tests the sensitivity of the solutions of the k-median model to the shape of the underlying communication network in the case of autarky. More precisely, the k-median problem is applied to different squared and hexagonal toy-networks representing various transportation networks (Euclidean, quadratic, radial and circumradial). This chapter shows that there is a clear relationship between the shape of the network and the optimal locations and allocations and that the location of the peripheral link in circumradial configurations is of great importance to the efficiency of the solutions.

Chapter 4 extends the preceding simulations to other toy-networks and to another location model: the simple plant location problem (SPLP). Two types of economies are considered: the case of autarky as a benchmark and the case of two economies characterised by a different network configuration that forms a common market. The results obtained in Chapter 3 are confirmed: the location-allocation results highly depend on the shape of the transportation network; it is also shown that the structural properties of the transportation networks still hold in a common market (history matters), though some locations are pulled toward the common border.

In Chapter 5, we again consider a common market made out of two squared lattices of points, but this time the design of the transportation networks is given and fixed in both regional economies. Here, we consider the permeability of the border between the two entities (number of links, location of the links, length of the link) and also analyse the effect of the adjunction of a high-speed link between the regional centres of both economies on the location-allocation results. Our simulations reveal that the location and the number of gates have little impact on the optimal locations and allocations, but that the 'length' of the gate between the two regions (taxes, congestion) as well as the existence of a high speed link (such as high speed train, motorway) between the two regional centres influence the locational pattern. However their effect is hardly unpredictable, confirming the scepticism of some transportation analysts.

Part III: Many facility location models assume that customers select the facility they will patronise on the basis of distance. Hence, measuring the length of the edges of a transportation network is a time-consuming task that is of prime importance in most spatial analyses. When objects in space are represented by points, distance-predicting functions may be used to transform the point coordinate differences into an estimate of the travel distance between them. In this part of the book, the shape of the transportation network is given; it is the weight associated to the edges that varies.

Chapter 6 uses l_p distances to measure the length of the links of the transportation network; p varies from 1.0 to 9.0. The objective of this chapter is to test how robust the solutions of a location-allocation model are to changes in the parameters of a distance-predicting function and consequently with what accuracy these parameters should be estimated in real-world applications. This topic is interesting in GIS developments. The effects of systematic changes in p in the l_p norm are systematically analysed on the k-median results, using regular as well as irregular lattices of points. Different values of the rotation axes (θ) or of the exponents (α) are also applied to distance. All simulations lead to the following conclusion: changes in p, θ

and α influence the absolute values of the location parameters (such as the maximum or the average distance) but have little effect on the location-allocation results (the choice of the location sites).

In Chapter 7, the results of the UFLP model are compared for different price policies, on squared as well as on hexagonal networks. Using a price policy enables to take into account time, capital, work and the real cost of getting over the physical distance. Who pays for these costs: the consumer or the firm? What are the consequences of such choices on the optimal spatial organisation? Several price policies are used, showing the complex relations between geographical and economic distances. We show that the choice of a price policy has a limited impact on the location of firms compared to the transport network configuration. This conclusion points to the same direction as the results of Chapters 3 and 4. Although one has to be careful in extending conclusions from rather simplistic toy-networks to real-world situations, this suggests that shaping the transport network proves to be a rather powerful instrument in spatial planning, especially in countries with limited infrastructures like the less developed countries.

The preceding chapters are concerned with the effects of the measure of distance on location results (weights associated to the edges of the graph); it is quite obvious from our everyday experience, however, that there are other costs associated with distance that are not directly borne by those generating them. Hence, in Chapter 8, the optimal location is no more a 'simple' minimisation of the transportation costs: we consider the trade-off between transportation costs and the costs bound to the spread of a negative externality (such as pollution) produced punctually at the supply site and measured on discrete points on the network. It is shown that slight pollution does not modify the solutions of the location-allocation model. Important pollution generates competition between facilities and modifies their optimal locations; facilities are located further apart from each other and hence further from the consumers. Very important pollution tends to push the optimal locations to the border of the area under consideration. In other words, when the pollution is slight, accessibility of the supply sites remains the most important location criterion. When pollution gets more important, facilities tend to be located as far as possible from each other, leading to quite different spatial optimum.

The last part of this book (Part IV) is devoted to two real-world applications illustrating and confirming the results of the simulations performed on toy-networks in the preceding chapters. Chapter 9 considers the case of locating primary health care facilities in a province of Niger where travel is more difficult in the rainy season than in the dry season. We consider three types of transportation networks: the actual dry season network, the actual rainy season network and an enhanced rainy season transportation network. We show (and confirm) that the location results and hence the planners' decisions are sensitive to the measures of the transportation conditions. Chapter 10 illustrates the results obtained in Chapter 8: it considers the location of a community recycling centre within a Belgian residential 'commune'; it shows how far the optimal location and allocation patterns vary with increasing levels of noise pollution.

Chapter 11 concludes the book.

PART ONE

Measuring the shape of the transportation network

2. 'Measuring' the shape of a network: state of the art

Before going to the main topic of this book, it is interesting to understand the problem of 'measuring' the shape of the transportation network, that is to say: (1) of associating to one network one or several indices representing unequivocally the spatial arrangement of the nodes and edges, and (2) of comparing different graphs (transportation networks) and classifying them into groups of similar shapes. The final objective being to evaluate the impact of network morphology on the location of human activities, it would indeed be extremely useful if we could classify graphs by their morphology in order to better establish quantitatively the link with the location-allocation results. The results presented in this chapter were partially presented by Beguin and Thomas in 1997.

2.1 SOME DEFINITIONS

The geographical environment is by essence continuous. Practically, it is often represented by a network, a discrete set of nodes and edges limited by a border. The border determines the size and the shape of the studied area. Nodes often represent the locations of the (human) activities and have interaction potential. Nodes often summarise the information of a whole area: a node can, for instance, represent a household, a village, a township, a city, a firm or a train stop depending on the scale and on the objectives of the analysis. Raffestin (1991) and Dupuy (1985, 1994) speak about 'territorial nodosity'. This aggregation procedure highly biases the results of most interaction models (see for example Roy, 1999). Nodes are characterised by their number, by their geographical coordinates (their location) and by the weight and/or attributes associated to them. These last can, for instance, be a number of inhabitants, a number of square kilometres, a quantity of production, a total workforce, and so on, or other attributes related to the physical properties of the nodes or to the actors (see for example Peuquet, 1988; Kamann, 1998, p. 283). Hence, nodes are heterogeneous. Relations or exchanges are made possible by the existence of links between the nodes. In this study, nodes are given without any reference to the reasons of their existence.

The edges are the links between the nodes materialised in the real environment by highways, waterways, railroads, airways, pipelines, cables and so on. Edges in a network are characterised by their number, but also by their location, their length, their technical characteristics, the flows associated with them and their capacity. The design of the links depends upon the location of the nodes and on the characteristics of the environmental support of the links.

Hence, graphs are a common way of representing the geographical environment. As already mentioned, the words 'network', 'link' and 'node' used in applied sciences have the respective synonyms 'graph', 'edge' and 'vertex' in computer science; these terms are used here interchangeably. Motorway networks linking cities together or underground networks linking urban wards are common examples of such graphs. Transportation networks are the physical supports of many economic and transportation activities. Place of residence, workplace, location of firms, location of public services and even the geographical development of the cities themselves depend on accessibility and consequently, on the shape of the transportation network (see Anderson, Kanaroglou and Miller, 1996; Banister, 1995; Dijst and Vidakovic, 1997; Handy, 1997).

Spatial analysts considering a transportation system as one aspect of spatial organisation first look at the structure of the network. If, theoretically, all links between all nodes are possible (Figure 2.1, top), inspection of maps and related geostatistical materials show that some of the linkages are not possible because there is a natural barrier, or because the land is privately used, for example (Figure 2.1, centre and bottom). Moreover, some links can be considerably more important than others, leading to the ideas of channelisation (Hoover, 1975) and of hierarchy of the links. The most important linkages are identified as trunklines, corridors, main streets, arterials, channels or magistral lines (see Taaffe, Gauthier and O'Kelly, 1996, p. 5). The less important lines are called feeder lines. In some but not all networks, there is a third type of linkage, the bridge line, linking gateways. Bridge lines are designed to perform a transshipment function and to carry through traffic between two or more major transportation or regional systems (see Chapter 5). Hence, starting theoretically from a complete network where all linkages are possible between all nodes, we end up with a well-structured network, with a hierarchy of nodes and edges that would be interesting to describe by one or several indices. The problem is now how to measure this heterogeneity.

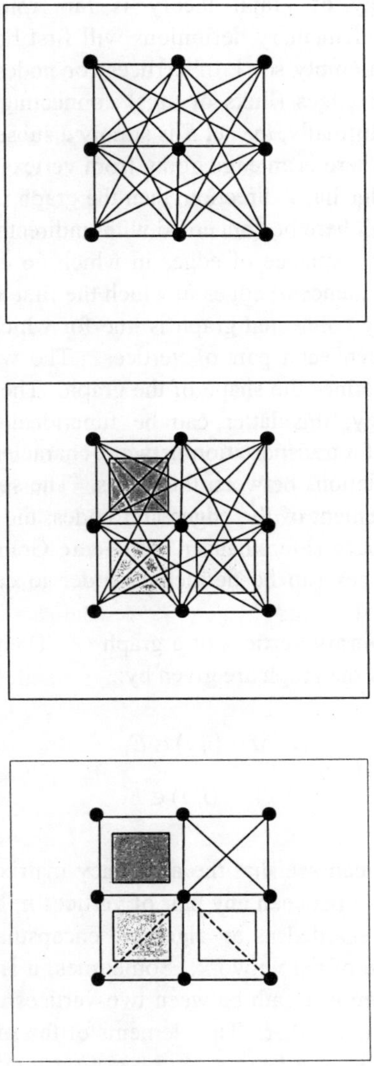

Figure 2.1 From a complete graph (top) to a structured transportation network (bottom)

Because the terminology of graph theory is far from being standard (Bollobas, 1990), some elementary definitions will first be summerised. A graph G is simply a non-empty set V of vertices (or nodes) together with a possibly empty set E of edges (links or arcs) connecting various pairs of elements from V. More formally, the set E is simply a subset of the Cartesian product $V \times V$. Thus, if there is an edge going from vertex i to vertex j, the pair $(i, j) \in E$. If each edge has a direction, then the graph is called a directed graph or digraph. We will here be concerned with undirected graphs only. A path is then defined as a sequence of edges in which no edge is used more than once; a cycle is a sequence of edges in which the first and the last vertex are the same. A strongly connected graph is one for which there is at least one 'simple' path between each pair of vertices. The way the nodes are connected in the space defines the shape of the graph. The shape of a graph relates to its morphology; this latter can be functional or spatial. The functional morphology of a transportation network characterises qualitatively and quantitatively the relations between the nodes. The spatial morphology shows the spatial arrangement of the edges and nodes, the way the network covers and drains the space (Frankhauser and Genre-Grandpierre, 1998, p. 66). Three useful matrices can be defined in order to capture the spatial structure of a graph.

Let i and j be two arbitrary vertices of a graph G. Then, the elements of the adjacency matrix A of the graph are given by:

$$A_{ij} = \begin{cases} 1 & \text{if} \quad (i, j) \in E, \\ 0 & \text{if} \quad (i, j) \notin E. \end{cases}$$

From this definition, we can see that the adjacency matrix A tells us when there is a direct connection between any pair of vertices in the graph G. The matrix A can hence be regarded as an algebraic encapsulation of the most basic connective structure of the network. Sometimes, it is more interesting to see whether or not there is a path between two vertices than if we can go directly from one vertex to another. The elements of the reachability matrix R of the graph are given by:

$$R_{ij} = \begin{cases} 1 & \text{if} \quad \text{there is some path} \quad \text{linking } i \text{ and } j, \\ 0 & \text{if} \quad \text{there is no path} \quad \text{between } i \text{ and } j. \end{cases}$$

We can then also define the element D_{ij} of the so-called distance matrix D as the length of the shortest path connecting the vertices i and j, where $i,j = 1, 2, \ldots, n$. Interestingly, this shortest path matrix is also one of the main input of location-allocation models. A direct relationship between the three mentioned matrices can easily be established (Casti, 1995, p. 7), but they are not sufficient to summarise unambiguously the spatial structure of a transportation network.

Now, suppose that we have a second graph G' and that we want to know whether the two graphs G and G' have the same connective structure. Clearly, this will be the case if and only if the two graphs differ only by a relabelling of their vertices. The graphs G and G' are said to be isomorphic if and only if there exists a $n \times n$ permutation matrix P such that $PA(G') = P^{1}A(G)$. Note that two isomorphic graphs can correspond to two different images in the Euclidean space. Hence, our interest in finding and using a method to detect isomorphism and to measure how far two networks look alike.

Further theoretical developments about graph theory can be found for instance in Berge (1963, 1973), Busacker and Saati (1963), Bollobas (1990), Casti (1995) or Wilson (1996). Instead of continuing in this (abstract) mathematical fashion, let us now talk about two practical problems and the methods available for their resolution: is it possible to characterise unequivocally the shape of a transportation network by one or several numerical indices? Is it possible to identify two isomorphic network structures?

2.2 NETWORK INDEXING IN THE LITERATURE

As already mentioned, our final objective being to evaluate the impact of network morphology on the location of human activities, it would be extremely useful if we could synthesise the morphology of a given graph by one index; this index should be conceptually descriptive, easily measurable and suitable at a variety of scales. According to Griffith et al. (1986), the ideal shape index should enable the regeneration of the original shape from the value of the index; additional properties include such characteristics as independence of scale, independence of the translation of the figure and/or the starting point, and independence of the rotation of the figure. Simply stated, the shape index should be independent of the coordinate system used in digitisation of the figure. An implicit assumption in shape analysis is that there is a one-to-one correspondence between the resulting index and the figure it represents. Hence, the first general concern of this book is to try to associate to each graph/network one or several indices in order to best describe the spatial organisation of the nodes and links (the shape of the network).

The literature devoted to network indexing is large. It relates to different disciplines and often dates from the 1960s and 1970s (Table 2.1). Numerous indices describing the shape of a network are directly inspired by graph theory (to name but a few: Berge, 1973; Kansky, 1963; Busacker and Saaty, 1965; Tinkler, 1977; and Richardson and Young, 1982). Kansky (1963) summarises several indices expressing the centrality, the connexity (the continuity of the network), the connectivity and the nodality of a network. Some of these indices are called global (for example among others the cyclomatic number, α, β, γ, η,), others local (for example Shimbel, diameter) (for example Berry and Marble, 1968; Dupuy and Stransky, 1996). In general, these indices are easily computable, but they do not take into account the length of the links, nor the quality of the communication, and they are not able to discriminate between networks of identical size (see for example James et al. 1970; Smith, 1975; Cliff, Haggett and Ord, 1979; Musso and Vuchic, 1988). This is mainly due to the fact that their computation is mainly based on the number of links (e) and/or nodes (n) and not on the spatial structure (morphology) of the graph. It has been shown that for a given number of edges and nodes, it is easy to find topologically different networks characterised by similar values of the descriptive indices. This was also true for the networks used in this book (see Arnold, 1994 for the networks used in Chapter 4 of this book). The application of the most commonly used indices in graph theory did not lead to a clear clustering of the transportation networks. These indices would be of no use if we want to associate (correlate) them with the quantifiable outputs of the location-allocation models.

Table 2.1 Literature about graph/network indexing

Type	Authors	Main indices
Graph theory	Berge (1963, 1973), Busacker and Saaty (1965), Kansky (1963), Garrison and Marble (1963), Pitts (1965)	Centrality (Königsnumber, accessibility indices), Morphology (indices β, γ, μ (cyclomatic number), α), Compacity of the network (δ diameter; Shimbel D)
	Examples of applications: Lowe and Moryades (1975), Taafe and Gauthier (1973), Musso and Vuhic (1988), Laporte, Mesa and Ortega (1994), Dupuy and Stransky (1996), Haynes (1977)	Critics: indices based on the number of nodes and links see Haggett, Cliff and Frey (1977), James et al. (1970), Smith (1975)

Type	Authors	Main indices
Shortest path matrix	Ord (1967), Haggett, Cliff and Frey (1977) Examples of applications: James et al. (1970), Cliff et al. (1975), Haggett and Chorley (1972), Cliff, Haggett and Ord (1979)	*S-I* indices Where *I* = variance/mean *S* = skewness/variance. Critics: discriminate types of distributions but not the shapes of the graphs Cliff, Haggett and Ord (1979)
Geomorphometry	Horton (1945), Strahler (1964), Shreve (1966, 1967), Scheidegger (1965), Jarvis (1977), Jarvis and Werrity (1975), Ferguson (1975), Liao and Scheidegger (1968), Werrity (1967)	Order, bifurcations, forks Critics: only valid for oriented trees (drainage basins); identical values of the indices for topologically different shapes Haggett and Chorley (1972)
Eigenvalues	Cliff and Ord (1981), Boots (1984, 1985), Boots and Royle (1991), Griffith et al. (1980, 1996), Haynes (1977), Gould (1967)	λ max Critics: empirical and intuitive; no clear indicator for the spatial structure of a graph Hay (1975), Tinkler (1972, 1975, 1977)
Fractals	Klinkenberg (1992), Batty and Longley (1994), Montgomery (1996), Frankhauser (1994, 1997), Benguigui and Daoud (1991), Benguigui (1995) Frankhauser and Genre-Grandpierre (1998), Goodchild and Mark (1987)	Critics: not capable of taking the entire complexity of the problem; sensitive to the way data are summarised Klinkenberg (1992), Klinkenberg and Goodchild (1992), Batty and Longley (1994)

Ord (1967) used two indices (*S* and *I*) to describe and classify networks; the computation of these indices is based on the frequency distribution of the distances associated with the shortest-path matrix between the nodes of the graph. Comparisons of *S* and *I* values are simply made by a bi-dimensional diagram. *S* and *I* depend on the moments of order 1 and 3. Planar graphs often refer to beta-binomial distributions, non-planar graphs refer to hypergeometrical distributions. When *e* increases, *I* tends to 0 and *S* to –1. Weights can be considered. It has already been demonstrated that these indices can be useful in describing real-world networks but they are not able to cluster them unequivocally: two pairs of identical *S-I* values may correspond in reality to different networks (see Cliff, Haggett and Ord, 1979).

Geomorphometry is the quantitative description of the geometrical features of the land surface; in practice, it is often limited to the description of rivers or drainage basins and therefore to graphs. Rivers are seen as planar graphs with sources, forks or bifurcations, and a mouth. Hence, the studied graphs are often limited to directed trees. Indicators are mainly based on the notion of 'order'. The definition of the order depends on the author and on the nature of the measure (nominal, ordinal, cardinal) (see for example Horton, 1945; Strahler, 1964; Scheidegger, 1965; Schreve, 1966, 1967; Chorley, 1972; or Mark, 1975). Werner and Smart (1973), Jarvis (1977) and Jarvis and Werrity (1975) classified rivers on the basis of several topological criteria. These methods are difficult to extrapolate to transportation networks, as these latter are most of the time bi-directional and not very often limited to trees. Moreover, it has been shown that two non-isomorphic networks can be characterised by identical values of the indicators; geometrically different networks in the Euclidean plane can be topologically identical and networks with a similar number of sources can have a similar topological complexity, but are topologically different. Most studies conclude that more morphometric measures are required in order to yield more robust signatures.

Another category of descriptors that is worth mentioning here is the eigenvalues and eigenvectors based on adjacency matrices described by Kendal (1957) or Cliff and Ord (1981). These authors conclude that a highly connected network generates large eigenvalues, the higher the valence of a link, the higher the eigenvector. This method has however been deeply discussed by Hay (1975), Tinkler (1972, 1977, 1979), Boots (1984, 1985), Boots and Royle (1991), Griffith and Jones (1980) and Griffith (1996) because different eigenvectors do not always conduct to a non-ambiguous interpretation (such as in factor analysis). This method is mainly intuitive and empirical; it has not lead to the construction of a clear indicator 'describing' the structure of a graph. Once again, different eigenvalues do not lead to an unambiguous interpretation.

More recently, several attempts have demonstrated the potential role that fractal parameters (Mandelbrot, 1977) may have as morphometric measures for rivers (see for example Klinkenberg, 1992; Montgomery, 1996), cities (Batty and Longley, 1994; Frankhauser, 1994, 1997), transportation systems (Benguigui and Daoud, 1991; Benguigui, 1995; Frankhauser and Genre-Grandpierre, 1998) or topography in general (Klinkenberg and Goodchild, 1992; Goodchild and Mark, 1987). The fractal approach is not based on the notion of density, but mainly on internal hierarchy. It has been shown that the fractal dimension captures some complementary aspects to the more 'classical' approaches of network morphology, but fractals are not yet capable of taking the entire complexity of the problem into account (Klinkenberg, 1992; Klinkenberg and Goodchild, 1992). Moreover, the fractal dimension is quite sensitive to the way data are summarised (Batty and Longley, 1994).

Last but not least, a very active and promising area of application is the field of pattern recognition (Attneave and Arnoult, 1966; Medda, Nijkamp and Rietveld, 1998; Fu and Bhargavan, 1973). These methods, however, mainly apply to surfaces, like urban outward form, and their use is far beyond the scope of this analysis. Among the large number of different applications, an interesting one appears to be the method developed by Lee in 1976 who examined the shapes of chromosomes in order to classify them into three categories. He used fuzzy pattern recognition, a technique that enables comparison of the angles and arm lengths of the chromosome with those labelled in the idealised chromosome shape. His method has the advantage of being insensitive to rotation, translation, expansion or contraction of the shape, but it is based on the assumption that there is one idealised shape pattern (or several). In transportation networks, unfortunately, this is not the case. It is therefore worth continuing our investigation into finding alternative and/or complementary approaches.

Computer sciences and neural networks modelling techniques are also new directions helping not only indexing, but also analysis through matching and synthetical modelling (Peura, 1997 and 1998). They seem to be quite promising techniques easily applied to simple networks although they quickly reach a very high level of complexity far beyond the simplicity of the indices used up to now by geographers or geomorphometists. Some indexing and matching techniques were used in order to compare simple (simplistic) and small-sized tree-shaped networks (Thomas, 2000, pp. 36–48). At the current stage of knowledge, their application to transportation networks such as those used in this research are still not possible as they would be much too time-consuming. We think that these methods provide quite promising techniques in order to 'measure' and classify the shapes of the transportation networks, although they quickly reach a high level of complexity, far beyond the simplicity of the indices used up to now by geographers or geomorphometrists and far beyond the objectives of the present approach.

To complete the review, let us also mention that several authors are concerned with the measurement of the transportation access level (see for example Morris, Dumble and Wigan, 1979; Allen, Liu and Singer, 1993; Pooler, 1995; or Gutiérrez, Monzon and Pinero, 1998). Traditional measures are the gamma index, the *T*-matrix and the *D*-matrix. The gamma index refers to the number of linkages in the network, the *T*-matrix is the total accessibility matrix and the *D*-matrix is the Shimbel distance matrix. The *T*-matrix is a measure of the total direct and indirect connections to a node on a network; the *D* matrix is a measure of the shortest path between nodes in a network (Taaffe, Gauthier and O'Kelly, 1996). This is, however, a problem different from ours as it does not refer to the spatial topological morphology of the transportation network

Despite all the mentioned scientific efforts, describing a graph/network unequivocally seems to be a dream for the geographers. Up to now, the existing literature confirms that, strictly speaking, it is theoretically impossible to describe a graph unambiguously with one or just a few descriptors, unless the graph is infinitely small (two nodes, one edge). The reason lies in the fact that a graph contains more information that can be transmitted by a (small) set of numbers. The information inherently equals the structure of the graph: the complete description of a graph is the graph itself. Thus any index is inevitably just a projection, scratching the surface of a wider truth. In general, the main problems with the existing measures are that indices are often highly correlated and that they may produce the same values for graphs exhibiting very different patterns (see Haggett, Cliff and Frey, 1977). This confirms what Griffith et al. wrote in 1986 about urban shape analysis: 'the failure to extract any one to one correspondence between a shape and its associated index value(s) continues to cast doubt on the utility for subsequent studies'.

Given the inability of the existing indices to capture unambiguously and easily the spatial morphology of a network, there will be no quantification of the spatial morphology of the transportation networks used in this book, no statistical test, no quantitative measure of the relationship between the results of location-allocation models and the spatial morphology of the transportation network. This book will be limited to non-statistical 'proofs'; it will simply illustrate the relationship by tables, figures or simple statistical material and interpret the results in terms of geographical common knowledge.

PART TWO

Location-allocation results and the shape of the transportation network

Introduction to Part II

Transportation infrastructure is often considered as one of the main instruments in the tool-kit of spatial planners (see for example Haggett and Chorley, 1972; Taaffe, Gauthier and O'Kelly, 1996). Many decision-makers interested in the role of transportation infrastructure take it for granted that more infrastructure is always better than less because it would lead to less congestion and/or to a higher accessibility to existing facilities. In fact, the relationships between the provision of transport infrastructure and the level and distribution of human activities is complex and can be addressed from different perspectives. First, one may wonder what is the link between the provision of infrastructure and the rate of economic growth? Second, one may also ask what is the influence of a transportation infrastructure on the attractiveness or accessibility of a given area in which the infrastructure is built up? In this part of the book, we focus on a third question: what is the impact (if any) of the shape of the transport infrastructure on the optimal geographical distribution of activities? To do so, we take the level of activity as given and study the geographical distribution of facilities within various exogeneously given network types.

As far as we know, this facet of the problem has not been much studied. The overwhelming majority of contributions disregard the impact of (re)shaping the transport system on the locational pattern of human activities. Either space is modelled as being one-dimensional or the shape of the network is given; both approaches do not permit assessment of the impact of the network shape (see Aoyagi and Okabe, 1993, for a related criticism). Yet, in facility location analysis, it has been shown that the choice of a particular metric to model distance leads to quite different locational models (Thisse, 1974; Beckmann and Thisse, 1986). Furthermore, the Hakimi theorem establishes that the search for a cost-minimising location along a network may be limited to the vertices of the network, thus showing that the facility location depends on where the nodes are (Handler and Mirchandani, 1979). These results are clear indications that the shape of the transport network is likely to have a significant impact on the optimal locations and allocations of facilities. Finally, the empirical evidence collected by Plassard (1976) about France and Italy reveals that new settlements arise only at the highway nodes, thus confirming what the Hakimi theorem suggests.

In this part of the book, two location-allocation techniques are used (the *k*-median and the SPLP). Simulations are performed on different toy-networks. The outputs of the models are systematically analysed in order to see how far the location-allocation operational results are sensitive to the shape of the transportation network.

Chapter 3 tests the sensitivity of the solutions of the *k*-median model to the shape of the underlying communication network in the case of autarky. More precisely, the optimal location model is applied to different toy-networks designed on a squared lattice of points (for example complete, radial, circumradial and quadratic transportation networks). All outputs of the location-allocation model are analysed separately and show: (1) that there is a clear relationship between the shape of the network and the optimal locations, and (2) that the location of the peripheral link in circumradial configurations is of great importance to the efficiency of the solutions. These results are confirmed by similar simulations performed on a hexagonal lattice of points.

Chapter 4 extends the preceding test to another model (the simple plant location problem) and to different toy-networks. Two types of economies are considered: first, the case of autarky (one squared lattice of points) and then, the case of two economies that form a common market, each economy being characterised by a different transportation network (two squared lattices of points, one with a grid network, the other with a radial network). It is seen that a grid network tends to foster a more dispersed pattern of activities, while the centre of a radial network acts as an attractor. The transportation network highly influences the location results. The structural properties of the individual networks still hold in a common market, though some locations are pulled toward the common border: history matters for locational patterns as far as transportation networks are concerned.

Chapter 5 is a continuation of Chapter 4 in that, given a common market with two economies characterised by their own transportation structure, we now look at the way the two economies are bound together: is the border characterised by many gates? Where are they located? Do these choices influence the location-allocation results? What happens if an extra high-speed connection is added between the two regional centres? We here limit ourselves to simulations on squared lattices of points on which a grid network has been designed. It is shown that the permeability of the border has little impact on the SPLP results, but that a high-speed connection can have important – but often unforeseeable – consequences on regional developments (see the example of the TGV in France).

It is our belief that obtaining general results regarding the impact of the transport network on the location of facilities is highly problematic because of the combinatorial nature of the problem, even in the case of toy-networks. However we believe that the kind of experimental approach followed in these three chapters leads to interesting results because it relies more on what we know from spatial microeconomics. In this respect, using toy-networks, we have been able to study the impact of the construction of additional links on the location of activities. This is especially relevant when several economies are merged in a common market because national transport policies tend to neglect the global impact of international connections.

3. Optimal locations and transportation networks: the case of autarky

3.1 INTRODUCTION

This chapter is devoted to a first set of sensitivity analyses: how far are the results of the *k*-median problem sensitive to changes in the shape of the transportation network? We here consider successively squared and hexagonal lattices of points on which several transportation networks are designed. The studied lattices of points have no connection with the other regions (case of autarky). Most results developed in this chapter have been published in *Transportation Science* (Peeters and Thomas, 1995) for the squared lattices, and in *Urban Systems* and *Flux* for the hexagonal toy-networks (Arnold, Peeters and Thomas, 1995; Arnold et al., 1997).

The formulation of a facility location problem heavily depends upon the nature of the set S of sites, which are considered as eligible for the facilities. In this regard, one generally classifies them into continuous models when S is a subset of the plane, discrete models when S is finite and network models when S is a topological graph representing a transportation system (see for example Hansen et al., 1987). To emphasise the difference in the formulation of those problems, Love, Morris and Wesolowsky (1988) use the terminology of site-generating and site-selecting models. Whatever the type of problem, the optimal pattern of facilities is also strongly related to the way distances between potential sites for facilities and clients' locations are measured.

Most of the investigations about the relationship between distance measurement and optimal location have been directed towards identifying a proper subset of S that contains an optimal solution to the original problem. For instance, let us recall that a polyhedral metric defines the distance between any pair of points as the length of a shortest path joining them and made up of segments parallel to a set of pre-specified directions; examples are the rectilinear, the oblilinear (Huriot and Perreur, 1973) and the one-infinity distance (Ward and Wendell, 1980) (see also Chapter 6 in this book). Then, it is well known that the search for an optimal solution of a continuous minisum problem with polyhedral norm can be restricted to a finite set of points of the plane. In the case of networks, theorems such as Hakimi's (1965) allow us to confine the search to the vertices of the network in the case of a minisum problem and to a finite set of points of the edges of the graph in the case of a minimax problem.

Another perspective is to evaluate the impact of the morphology of a communication network on the solutions of location problems. The importance of such investigations for land planning is obvious, since it may give useful insights when designing a transportation system (see for example Kansky, 1963; Haggett and Chorley, 1969). This can be illustrated by the study of Plassard (1977) on the development of the highway system in France and its consequences on the location of industries. In another domain, Yinger (1993) states that 'the role of the transportation system has not been recognised in empirical work in urban economics' (p. 329). He showed how assumptions about the transportation system are crucial to urban residential models. He therefore compared the results of an urban residential model applied on a dense network of right-angled streets to those provided by a dense network of radial streets. He showed that switching from a radial network to a street grid alters only one parameter of the equations of the standard urban model but leads to striking different spatial patterns. Daskin, Hurter and Van Buer (1993) introduced a joint facility location and network design model and showed that in some cases changing the underlying network may be a more cost-effective approach to improving service than adding facilities.

This chapter is oriented in a similar direction: it analyses the results of the *k*-median location model on systematically generated toy-networks and discusses the influence of the shape of the network upon the location results. Autonomous regional or national economies (the case of autarky) with no connection with surrounding regions are considered. In the next chapter (Chapter 4), the discussion is broadened to common markets and to another location model. In order to consolidate the conclusions, two lattices of points are considered: squared (Sections 3.2 and 3.3) and hexagonal (Section 3.4).

In this chapter, *k*-median problems are run on networks. Different aspects of the impact of the shape of the network are of interest. Firstly, we investigate the geographical distribution of the facilities, trying to identify the patterns, which are more likely to be produced by certain types of networks. For instance, which communication networks tend to produce uniformly distributed facilities and which favour concentrations in some subregions? Secondly, we compare the relative sizes of the facilities: when do they have more or less equal sizes and when does a 'natural hierarchy' arise? Thirdly, we study the supply of the service to the population. We compare the shape of the market areas and the distribution of the distances of the demand points to the closest facility, our purpose being to assess the influence of the shape of the network on the efficiency and the equity in the supply. A final point, not analysed in this contribution, would be to map the flows generated by the different solutions in order to appreciate the congestion in the network and the difficulties raised in case of failure of some arcs.

3.2 DEFINITION OF THE INPUTS

Let us start with simulations on toy-networks designed on a squared lattice of points. One set of points is chosen among all possible configurations: an 18×18 squared lattice of points in which each point has the same spatial environment as every other point, with the exception of the points located on the border. Hence 324 points are considered, scattered regularly in the squared space. The size is governed by empirical results: it seems to be the best compromise between feasibility, reasonable computing time and practical results. The middle point of the lattice is added to these 324 points in order to make a central location feasible. This addition does not affect the general results mentioned hereafter; it just facilitates the interpretation and the comparison of the results.

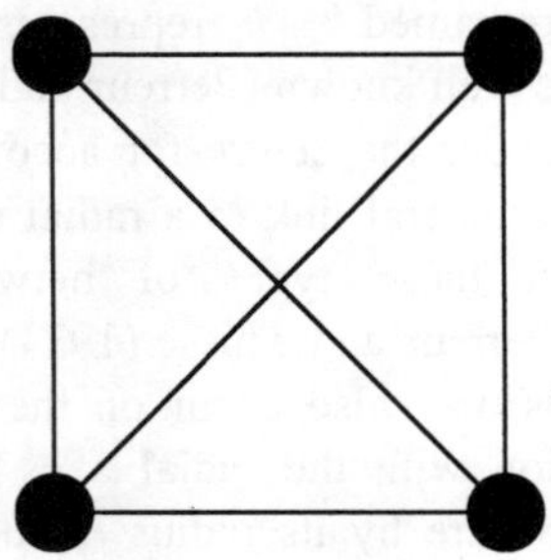

Figure 3.1 One fundamental cell of network a

Each of the 325 points of the network is a demand point as well as a potential supply site. In order to better isolate the tested effect, two sources of variation are kept constant: there is no spatial variation of the amount of demand allocated to each point (this assumption will be alleviated in the next chapter) and all demand/supply points are equidistant (the links between the nodes have the same length and are not directed). One exception is the middle point, which is located at the centre of the central cell. The 18×18 network is composed of 289 fundamental cells (Figure 3.1).
Several types of links and shapes of networks are designed on this 18×18 squared lattice of points. The first network, which will be used throughout as

a reference for comparing the alternative solutions obtained with the other networks, corresponds to a complete graph where distances are measured by the Euclidean distance between all pairs of nodes; it is denoted *Eucl.*

The other networks are represented schematically in Figure 3.2. Regular networks are named *a* to *f*, radial networks h_r, where $r = 0,1,..., 9$. *a* is a dense network with all vertical, horizontal and oblique (left and right) possible links; vertical and horizontal links have a length of 100, diagonals being set equal to 100 or 141 (see discussion below). Networks *b*, *c* and *d* are less dense: diagonals are alternatively (*b* and *c*) or totally (*d*) suppressed. *b* and *c* are mirror networks, both kept in the simulations with the hope that they would produce alternative optima. In network *d*, all diagonals are suppressed, leading to the rectilinear distance.

Radial networks are less dense (fewer links) than regular ones. In order to tend towards a comparable number of links, networks *e* and *f* are regular networks of the above-mentioned type *d*, where a number of links are randomly suppressed. Network *e* has 90 per cent of the links of network *d* and network *f* 80 per cent. Links are randomly deleted; *e* and *f* are asymmetric networks. The same *e* and *f* selections were kept through the analysis. When more than 20 per cent of the links are eliminated, the resulting networks are generally not connected anymore.

Circumradial networks are named h_r. h_0 represents a pure radial network, with no peripheral link. It is well known (Perreur and Thisse, 1974) that this type of network tends to favour the centre for accommodating an optimal location. When adding a peripheral link to a radial network, one obtains a circumradial network. In those types of networks, accessibility is theoretically increased and Perreur and Thisse (1974) show that the optimal locations of the supply sites may also occur on the peripheral road, more particularly at its intersection with the radial axes. The position of the peripheral road (represented here by its radius *r*) should thus be of prime importance on the location results. In this 18×18 network, *r* varies from 9 (location of the peripheral road near the centre of the network) to 1 (location of the peripheral road at the outer limit of the network, at the border). Network h_5 is represented in Figure 3.3.

In the networks mentioned above, diagonals of the regular networks and radial axes of the star-shaped networks have a length equal to 100. Following the Pythagorean rule, a length of 141 is also reasonable. Two sets of simulations are therefore considered in the following paragraphs: one with length 100 for the diagonals, following the assumption of equidistant points defined above, another with length 141, following the Euclidean properties of the space (Figure 3.1). Considering all networks from *a* to h_r means reducing regularly the number of links from almost 1200 (*a*) to 324 (h_0). This reduction partially explains some of the results discussed hereafter.

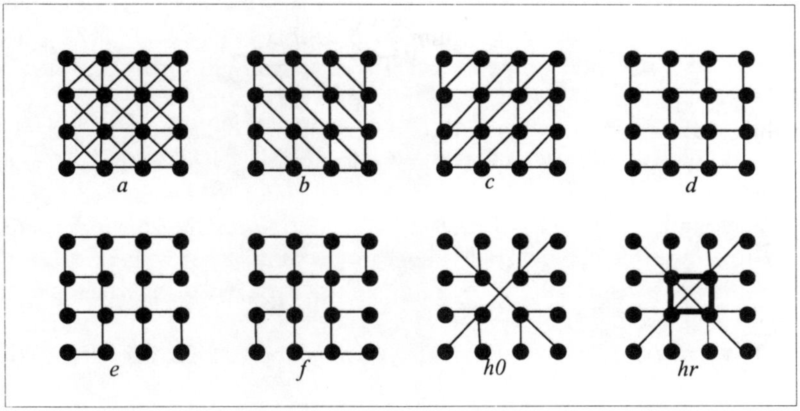

Figure 3.2 Schematic representation of the types of networks

For each experiment, the same location-allocation model is applied: the *k*-median model (see for example Hansen et al., 1987; Mirchandrani, 1990; Labbé, Peeters and Thisse, 1995, pp. 571–576; Labbé and Louveaux, 1997). The model was first enunciated by Hakimi (1964) and was subsequently shown to be *NP*-complete (Kariv and Hakimi, 1979). Heuristic and exact procedures to solve the problem are well known (see Teitz and Bart, 1968; Rosing et al., 1979; Hanjoul and Peeters, 1985; Densham and Rushton, 1992; Horn, 1996). The algorithm written by Hanjoul and Peeters (1985) will be used here. The *k*-median yields a configuration of *k* supply sites such that the sum of the distances between the demand points and their respective closest supply site is minimised. The output from a *k*-median model is a set of locations and their associated service areas. There is a long tradition of applications of such a location-allocation model to several types of public services (see for instance Laurent and Thomas, 1998 for a list of references).

The *k*-median model locates simultaneously *k* facilities such that the average distance from the users to their closest facility is minimised (constraint 3.1). It refers to a set of demand points (noted *i*) and a (possibly identical) set of candidate locations (noted *j*), all located at vertices of a network. It is assumed that each demand point is fully served by the server located closest to it (3.4). The number of facilities is one of the inputs of the model (3.3).

$$\min \sum_i \sum_j a_i d_{ij} x_{ij} \tag{3.1}$$

Subject to

$$\sum_{j=1}^{n} x_{ij} = 1, \qquad i = 1,\ldots,n \tag{3.2}$$

$$\sum_{j=1}^{n} y_j = k, \tag{3.3}$$

$$y_j \in \{0,1\} \qquad j = 1 \ldots n \tag{3.4}$$

$$0 \le x_{ij} \le y_j, \qquad i = 1,\ldots,n \qquad j = 1,\ldots,n \tag{3.5}$$

where

x_{ij} = fraction of the demand in i and satisfied by the facility in j (if any);
a_i = demand at node i;
d_{ij} = travel distance from node i to node j;
y_j = 1 if a facility is located in j and 0 otherwise.

The constraint (3.2) means that the demand in i must be met by some facility ; the constraint (3.5) prevents an assignment to a site that doesn't exist and constraint (3.3) ensures that exactly k facilities are located.

The model has here been chosen for its high flexibility, the good performance of the algorithms designed to solve it and its common use in location-allocation applications. Once again, in order to best isolate the tested effect, another source of variation is controlled: in this set of simulations, k is kept constant and set equal to 9. This constraint is relaxed in Section 3.3.7. Exact solutions are computed. Solutions are considered as erratic when execution time exceeds 3200 seconds.

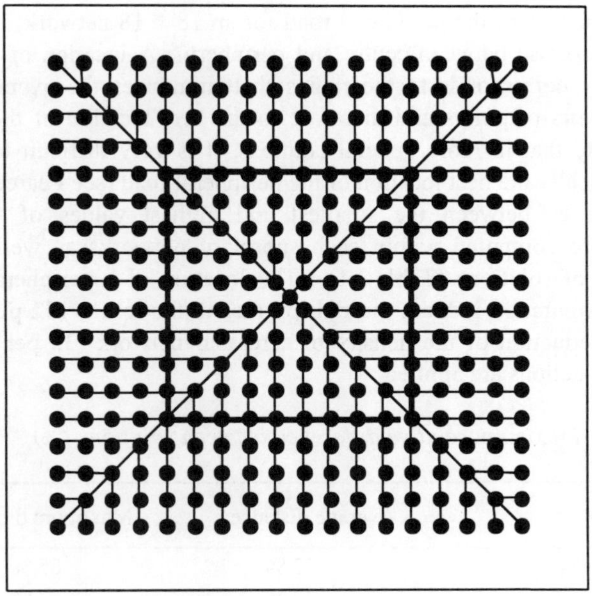

Figure 3.3 Example of circumradial network: h_5

3.3 RESULTS OF THE SIMULATIONS

In order to measure the influence of the shape of the network on the optimal location-allocation results, each output of the *k*-median will be discussed separately: the average distance (Section 3.3.1), the maximum distance (Section 3.3.2), the total allocation (Section 3.3.3), the location of the optimal supply sites (Section 3.3.4), the shape of the market areas (Section 3.3.5) and the distribution of the distance frequencies (Section 3.3.6).

3.3.1 Average Distance

Lets us consider the average distance between each demand point and its closest optimal supply site. Figure 3.4 shows that for regular networks (*Eucl*, *a* to *f*) the average distance increases regularly when switching from *Eucl* to *f*. Going to radial networks (h_0 to h_9), high values are observed for networks with no peripheral link (h_0) and for networks with peripheral links at the outer limits (h_9) with a minimum value observed for *r* equal to 3 or 4. h_3 and h_4 are the most efficient solutions for radial networks: they correspond to the optimal location of the peripheral road for an 18 × 18 network, that is to the best compromise between centre and periphery. A location of *r* in 3 or 4, empirically corresponds to the radius that minimises the average distance between pairs of points and therefore to the best location of the peripheral road. Note that in more general contexts, it is very difficult to determine mathematically the best location of the peripheral road (see Pearce, 1974).

Differences between the smallest and highest values of the average distance are computed within each subset of networks as well as for the whole set of solutions (Table 3.1). The location of a peripheral link on a radial configuration induces more disparities in the results (42 per cent) than does the reduction of the density of a regular network (32 per cent) when several reductions are applied.

Table 3.1 Maximum relative difference between solutions (%)

	Average distance	Maximum distance
Regular networks	32 %	65 %
Radial networks	42 %	74 %
Total set	72 %	106 %

Note: % = 100*(Max-Min)/Max

As shown in Figure 3.4, there is little effect of the length of the diagonal on the results: on average, the difference between the two sets of average distances is less than 9 per cent. Note that a dense regular network *a* with a diagonal length of 100 gives a lower average distance than does the Euclidean network *Eucl.* This can be explained by the metric used. A bold dotted line is used for networks *d*, *e* and *f* because they do not have any diagonal elements.

Figure 3.5 shows the average distance as a function of the number of links present in the studied networks: the regular relation is valid for almost all networks except h_r where *r* is greater than 1 and lower than 6 and especially when *r* is equal to 3 or 4. As previously mentioned, these two latter networks seem to correspond to the best trade-off between investment in the network and the quality of the service provided (and/or investment in facilities).

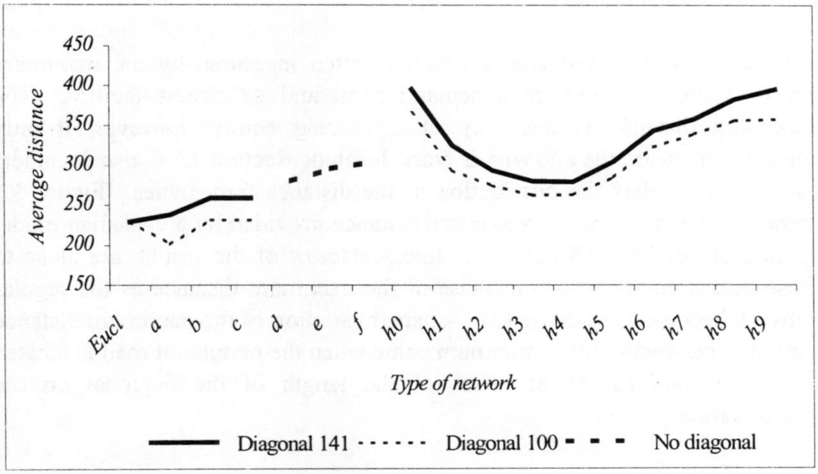

Figure 3.4 Average distance and type of network

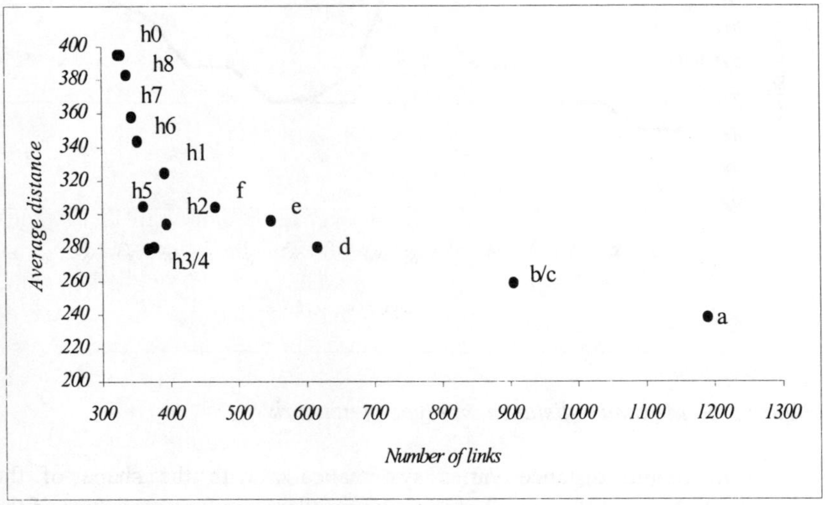

Figure 3.5 Average distance and number of links

3.3.2 Maximum Distance

The equity of a spatial configuration is often measured by the maximum observed distance between a demand point and its closest facility. The maximum distance is one way of measuring equity; however, it only concerns one user, the one who is worst located. Section 3.3.6 also considers equity: it considers the distribution of the distance frequencies. Figure 3.6 shows the variation of the maximum distance provided by a *9*-median model applied on an 18 × 18 network. Interpretations of the results are close to those discussed above: (1) increase of the maximum distance as the regular network becomes less dense (2) *U*-shaped variation of the maximum distance for radial networks with a minimum value when the peripheral road is located in $r = 3$; and (3) slight effect of the length of the diagonal on the interpretation.

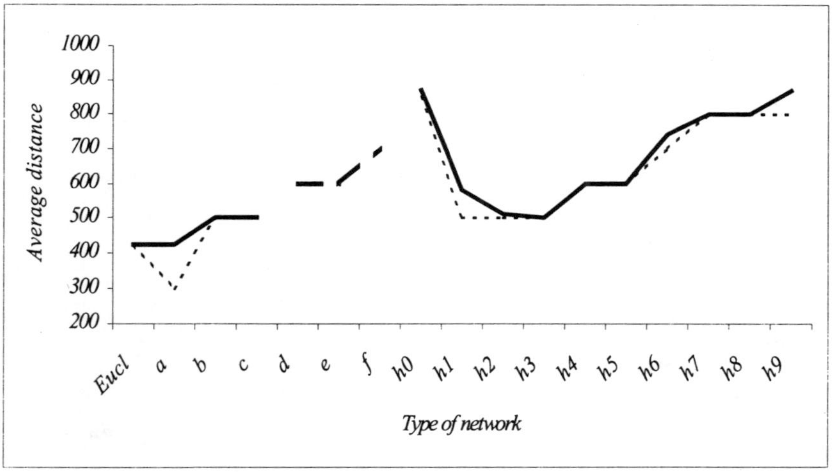

Figure 3.6 Maximum distance and type of network

The maximum distance varies systematically with the shape of the network. Relative differences between the smallest and highest values of the maximum distance are computed for each subset of networks and for the whole set of solutions: equity varies more than efficiency through the set of selected networks and the difference is higher within radial networks than within regular networks (Table 3.1).

h_3 corresponds to the 'best' location for the peripheral link: the maximum distance associated to h_3 is closer to rectilinear solutions than to other radial solutions. This location of the peripheral road corresponds to the optimal peripheral radius. This result has important implications for further empirical work on location-allocation models: planners should be aware that a non-optimal location of the peripheral link leads to a decrease in efficiency (Section 3.3.1) and a decrease in equity (Section 3.3.2).

3.3.3 Allocation

Figure 3.7 shows, for each type of network, the average, minimum and maximum allocation (number of demand points assigned to a facility) associated with the nine optimal supply sites; it gives a rough idea of the distribution of the demand associated with each pattern of optimal facilities. Small differences between the minimum and maximum allocation correspond to a balance between supply sites; large differences should not be chosen by the decision-maker especially when the service is not hierarchically organised. Figure 3.7 shows that some networks naturally lead to a more balanced configuration (*Eucl, a, b, c,* h_4) than others (*d, g,* h_1, h_2, h_3, $h_{5\ to\ 9}$), illustrating, once again, the danger if no control on the shape of the space is made. These results are not affected by the change in the length of diagonals (100/141).

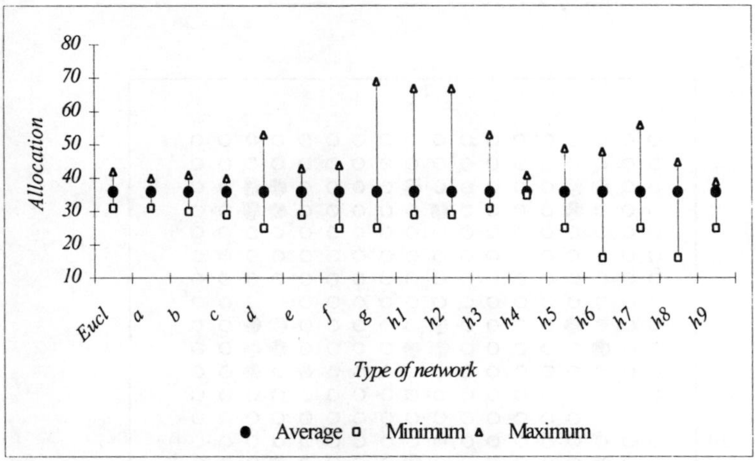

Figure 3.7 Minimum, average and maximum allocation

3.3.4 Location of the Supply Sites

Figures 3.8 and 3.9 summarise the optimal locations of the facilities within the various shapes of networks. For regular networks (Figure 3.8), optimal locations are regularly spaced locations with some slight differences from one network to another, hence defining 'optimal zones' for location. When comparing all 17 studied optimal solutions to this configuration, an index of dissimilarity can easily be computed. Hence Table 3.2 gives the number of times the optimal locations are located out of the 'optimal zones' suggested in Figure 3.8. Regular networks (*Eucl*, *a* to *d*) and radial networks with $r = 3$ or $r = 4$ (h_3, h_4) lead to equal solutions.

Table 3.2 Number of supply points falling out of the 9 'optimal zones' defined in Figure 3.8, for each network configuration

Eucl	*a*	*b*	*c*	*d*	*e*	*f*
0	0	0	0	0	5	6

h_0	h_1	h_2	h_3	h_4	h_5	h_6	h_7	h_8	h_9
4	6	7	0	0	8	7	5	7	4

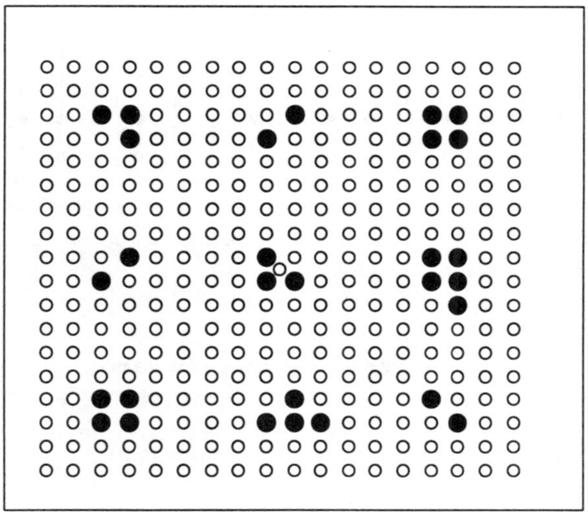

Figure 3.8 Optimal locations on regular networks Eucl, a to f

On radial networks, one optimal location is at the centre of the lattice, the others being often located on the peripheral link and more particularly at the junction of a radial axis and the peripheral road, confirming preceding theories (Perreur and Thisse, 1974). The location of the ring road leads to quite regular optimal locations except for large values of r.

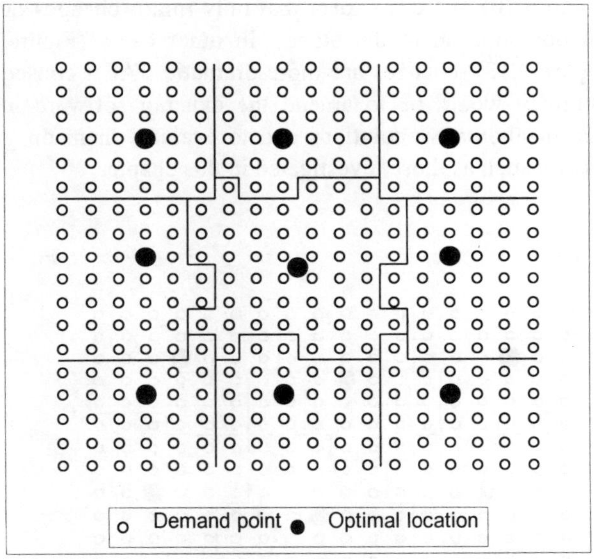

Figure 3.9 Optimal locations and allocations for the regular network a

3.3.5 Shapes of the Market Areas

Each demand point is allocated to its closest optimal supply site. Each supply point is therefore characterised by a zone including all allocated demand points (service or market area). For an emergency service such as the police (Thomas, 1993), an intervention team will most of the time work within well-defined district limits which are rarely passed beyond. Given the results put forward in the preceding paragraphs, the design of the optimal zones within the 17 spatial configurations is not surprising: regular networks are associated with a regular partitioning of the studied space (Figure 3.9). When a peripheral road is located in position 3 (or 4), on a radial network, the results are quite similar (Figure 3.10). However when r is different from 3 or 4, quite irregular and different solutions are suggested by the *k*-median model (Figures 3.11 and 3.12). Switching from a regular network to a radial network has little effect on the shape of the market areas when a peripheral road is added to the radial network and optimally located. Spatial consequences are very important in all other cases.

At this point, one should remind that symmetric networks typically produce many alternative optima. Therefore, the location patterns and the districting proposed in the various figures are not unique. For instance, any rotation of a multiple of 90° yields an optimal solution. In some cases (Figures 3.9 and 3.10), we can expect that only minor changes occur when passing from one solution to the other. In other cases (Figure 3.11 and especially Figure 3.12), changes are more dramatic. As a consequence, an interesting strategy would be to amend the existing software in order to generate a list of alternative solutions and to evaluate them on grounds of secondary criteria such as those investigated in this chapter.

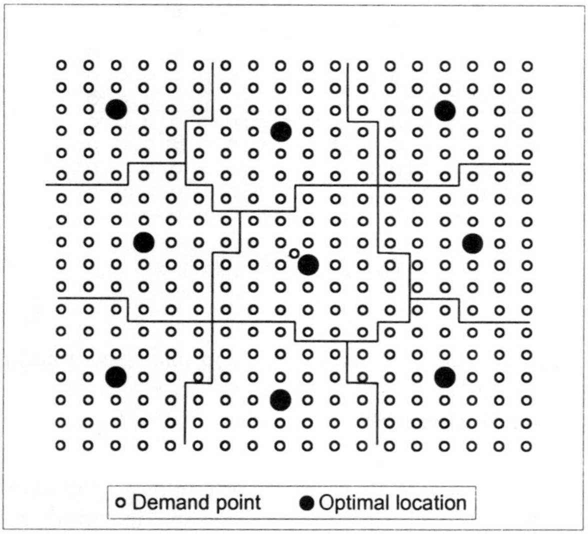

Figure 3.10 Optimal locations and allocations for the radial network h_4

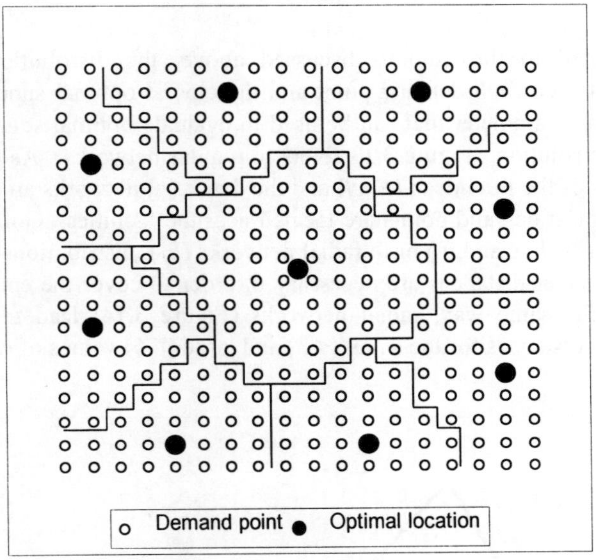

Figure 3.11 Optimal locations and allocations for the radial network h_2

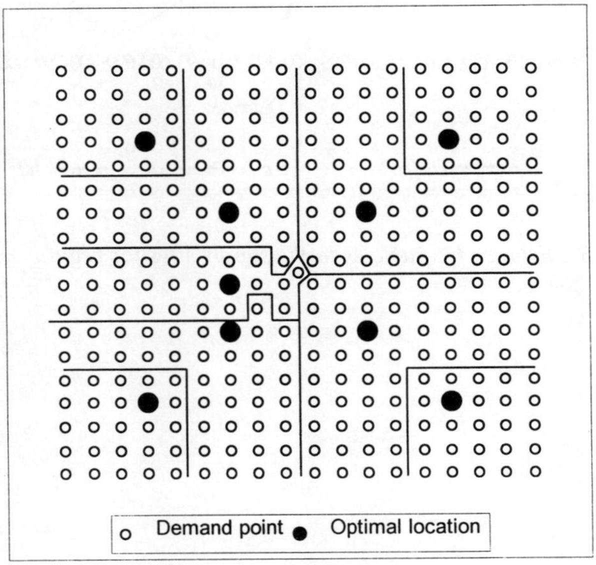

Figure 3.12 Optimal locations and allocations for the radial network h_7

3.3.6 Distribution of Distance Frequencies

Closely related to the outputs discussed above, the distribution of the distances between each demand point and the closest optimal supply site is another simple parameter that can be used to evaluate optimal solutions and equity of the solution. Figure 3.13 refers to regular networks. As expected, the distances of the Euclidean (*Eucl*) and the dense (*a*) networks are similarly distributed: most demand points are located near the facilities. Going to less dense networks (*b*, *c* and *d*) or to radial networks (h_r), distributions are more spread and longer distances are necessary in order to cover the entire set of points. In the same way, radial networks (Figure 3.14) lead to a larger dispersion of distances for the small (< 3) and large (> 5) values of *r*.

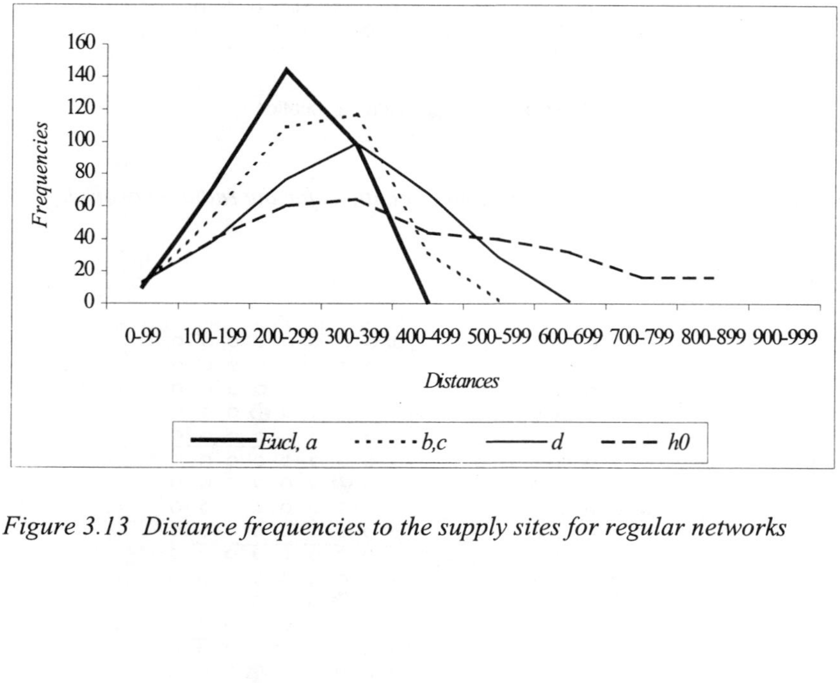

Figure 3.13 Distance frequencies to the supply sites for regular networks

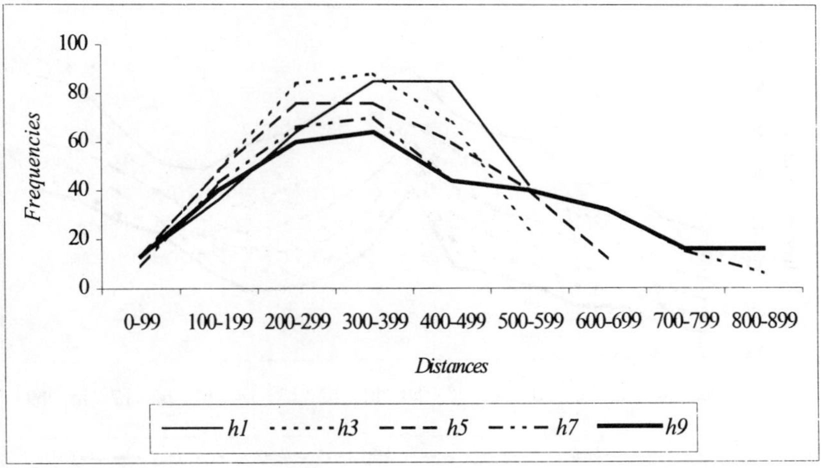

Figure 3.14 Distance frequencies to the supply sites for circumradial networks

3.3.7 Comparison with Other Values of *k*

Many sources of variation have been kept constant in order to isolate the tested impact of the morphology of the network. In this section, we change the number of facilities to be located; in the next section, the shape of the lattice of points will be modified. Figure 3.15 relaxes the constraint that fixes $k = 9$ and shows that the observed regularities are not sensitive to the investigated range of k: regular increase of the average distance when regular networks become less dense and a *U*-shaped variation for radial networks. However when k decreases, the optimal location of the peripheral road changes and r increases (the optimal location of the ring road is closer to the centre).

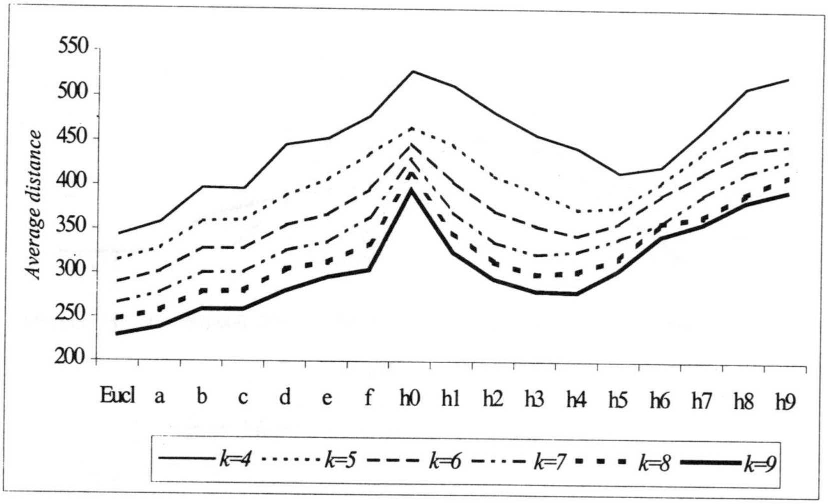

Figure 3.15 Variation of the average distance with the type of network and the number (k) of facilities

3.3.8 Conclusion

The main conclusion of this section is that the shape of the network has indeed an influence on the optimal solutions. This influence can be observed in different manners. Firstly, the optimal patterns of the supply sites appear to be closely related to the family of networks. Secondly, it turned out that dramatic variations in the shape of the service areas (that is, the zone allocated to each facility) could occur. By consequence, the workloads of the various facilities exhibit strong disparities for certain classes of networks, leading to a hierarchy of centres. Thirdly, the morphology of the network was also shown to have a marked impact on the differences in accessibility to the service between clients, being thus a source of unequity among the latter.

In the following section, we propose to verify these results by considering a hexagonal lattice of points. Three arguments claim for such a choice. Firstly, one may argue that since the nodes of the networks considered in the preceding section arose from a specific lattice of points in the plane while their edges were taken as proportional to the Euclidean distance, then, whatever the selection of edges used to construct a particular network, the relative positions of the points on the lattice will affect the measure of the distances. Therefore, we want to assess the robustness of the observations for the square lattice by considering a different pattern of points. Secondly, a hexagonal lattice is often considered in theoretical geography, for instance in the Christaller–Lösch theory of urban systems. In this respect, hexagonal lattices are much more common than square lattices in theoretical geography. Thirdly, hexagonal lattices allow one to construct networks that are quite different from those defined on a square lattice.

3.4 EXTENSION TO HEXAGONAL LATTICES

3.4.1 Designing Several Transportation Networks

Twelve different transportation networks have been designed on a hexagonal lattice of 343 points regularly spaced; the lattice is made out of 612 elementary cells, each cell being an equilateral triangle whose edge length equals 100. The shape of the lattice follows the marketing principle of Christaller central place theory (see Alao et al., 1977; King, 1984). Most results presented here are drawn from Arnold (1994), Arnold et al. (1995) and Arnold et al. (1997).

Regular networks were first generated. Network *A* is obtained by joining each point to its closest neighbours by an undirected edge; note that network *A* is not complete and is characterised by 954 edges. This network will be used as a reference throughout this section for comparing the solutions of a location problem defined on it with the solutions obtained using other toy-networks. Network *B* is similar to *A*, except that all the edges oriented in a given direction of the hexagonal lattice are deleted. It should be noted that even if there are three possible directions for deletion to obtain *B*, all yield to the same result, which is topologically equivalent to a rectangular configuration. *B* has 636 edges (Figure 3.16).

Asymmetric networks are then considered. They are also obtained by deletion of edges from network *A* and are denoted by C_d and D_d, where d varies from 1 to 3 and refers to the number of directions in which deletions are performed. These families of networks correspond to network *A* with respectively a regular (*C*-type) or a random (*D*-type) deletion of edges.

Radial networks were generated among which four are represented in Figure 3.16. These networks privilege paths going through the centre. Radial networks differ from each other according to the number of radial axes (3 or 6) and to the organisation of the secondary axes. They all have one peripheral ring road; its radius r varies from 0 (no ring road) to 9 (ring road located the furthest from the centre of the hexagon). This makes ten networks for each of the three following classes: $R3_r$ has three radial axes collecting secondary roads (342 to 396 edges); $R6_r$ has six radial axes collecting secondary roads (342 to 396 edges); RH_r has six radial axes collecting more hierarchically the secondary roads (342 to 396 edges).

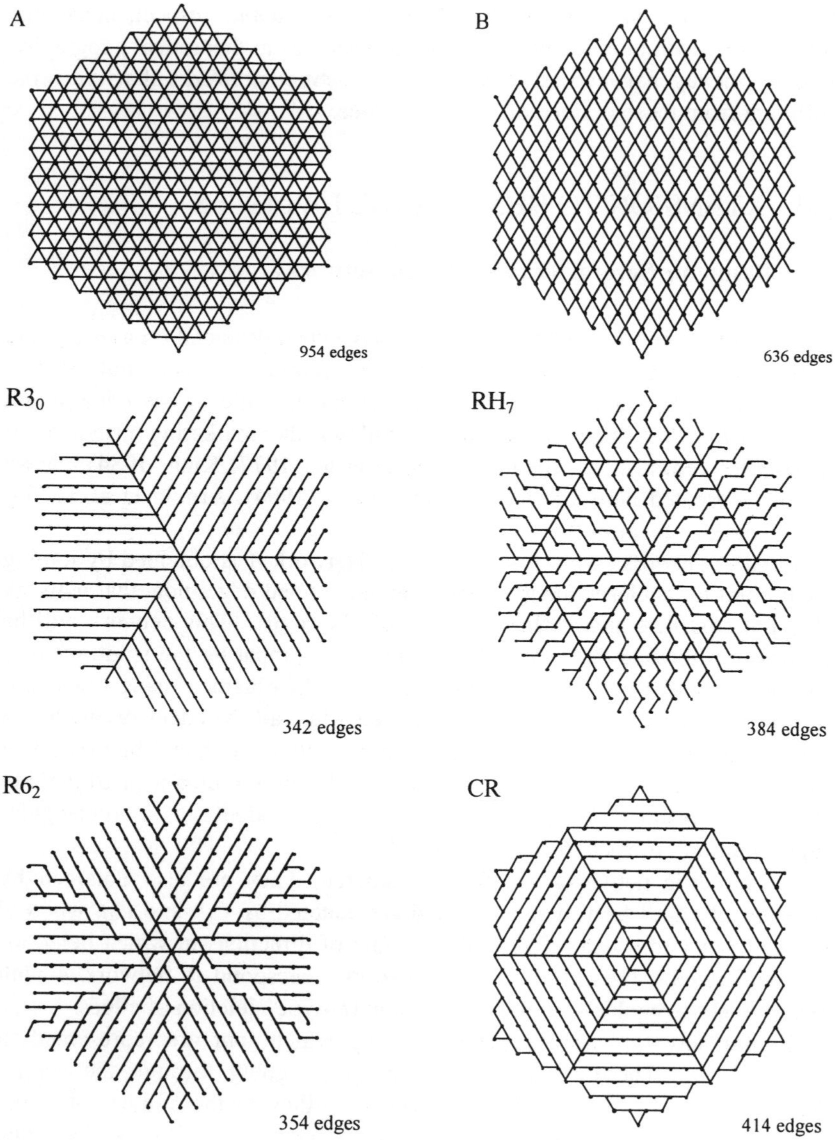

Source: Arnold (1994)

Figure 3.16 Regular networks (A, B) and examples of radial networks

Finally, *CR* corresponds to a radio-concentric network (Figure 3.16) with six radial axes and six peripherals; it has 414 edges.

These 31 selected networks constitute a sample of all the possible network shapes that can be potentially designed on the selected hexagonal lattice of 343 points. A 7-median model has been applied on each of them; the choice of the value 7 is justified by the hexagonal geometry: six equilateral triangles, each of them having an apex at the centre of the hexagon. The results are discussed in the next section.

3.4.2 Sensitivity of the *k*-median Results to the Transportation Structure

Similarly to Section 3.3, each output of the *k*-median is briefly but separately discussed: the distances between demand and supply points, the locations of the optimal supply sites and the total allocation as well as the shape of the market areas. We here only refer to some of the many simulations performed (see Arnold, 1994 for further details).

Distances between demand and supply points

The average distance between supply and demand (efficiency) and the maximal distance (equity) vary similarly with the characteristics of the networks; hence we will here only comment on the case of the average distance. Figure 3.17 shows the variations of the average distance between demand and supply with the number of edges and the type of network. Figure 3.18 shows the variation of the average distance with the position of the ring road in radial networks ($R3_r$, $R6_r$ and RH_r). These figures show that (1) as expected, the average distance decreases when the number of edges increases, that is to say when the connectivity of the network increases; hence the denser the network, the lowest the average distance. (2) All radial networks have the same number of edges for a given value of r but the computed average distances are different; this shows that the spatial structure of the transportation network can in itself modify the efficiency and the equity of the *k*-median solution. (3) The location of the ring road influences efficiency and equity: they increase with r until a distance very close to the outer limits of the hexagon. However it is impossible to separate the effect of the location of the ring road from the effect of the density of edges because the number of edges increases automatically with r. (4) Finally, when the ring road is located far away from the centre (r = 7 or 8), the value of the average distance turns around 300 or 500, that is to say the same value as for non-radial networks that are characterised by a much large number of edges (almost 50 per cent more); this is another indication that the spatial structure of the transportation systems matters.

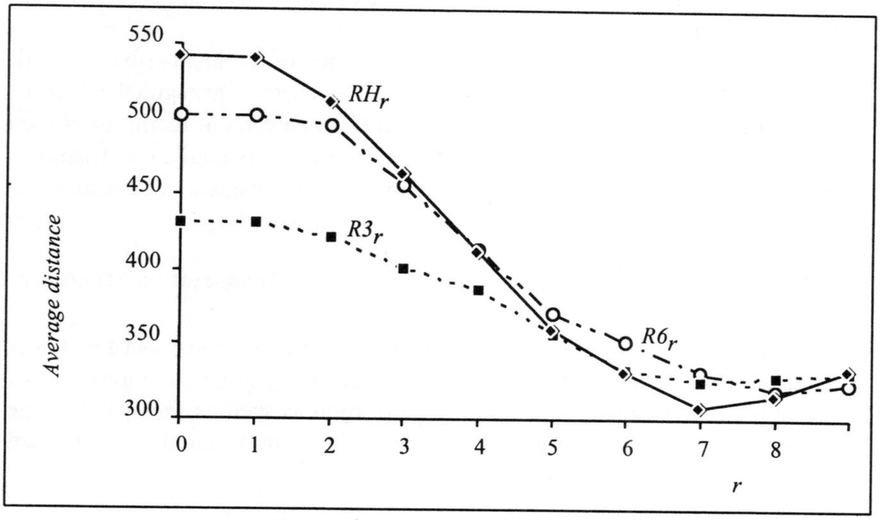

Source: Arnold (1994)

Figure 3.17 Average distance and the number of edges

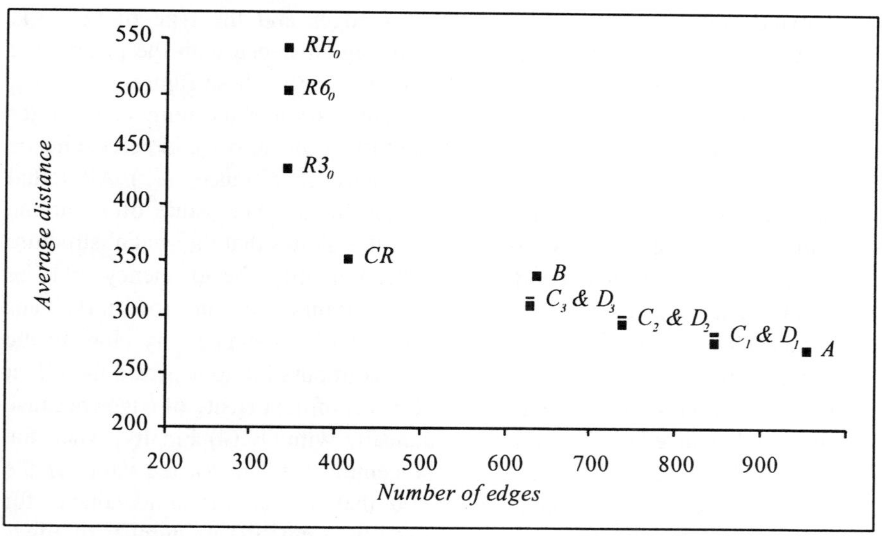

Source: Arnold (1994)

Figure 3.18 Average distance and the location of the ring road

The distribution of the distances between each demand point and its closest optimal supply site is another way evaluate the optimal solutions of a location-allocation model. The results obtained on hexagonal networks are similar to those obtained previously on squared lattices. Network *A* leads to a bell-shaped distribution, symmetric around the mean (see Arnold, 1994 for further illustrations). When the number of edges is smaller (C_d and D_d), the distribution of the distances tends to be asymmetric and less peaked. Results are similar in the case of regular (C_d) or random (D_d) deletions of edges. The comparison of the distributions of distances for several types of network (*A*, *B*, *CR*, $R3_0$, $R6_0$, RH_0) show that radial networks generate more widespread distributions than the *A*-type network and thus induce more disparities among clients. Finally, it appears that in the case of regular and asymmetric networks (*A*, *B*, C_d and D_d) demand points are located near the facilities, while for radial networks ($R3_0$, $R6_0$ and RH_0) longer distances are observed, the radio-concentric network (*CR*) giving intermediate values.

Sensitivity of the optimal locations

Figure 3.19 summarises the 7 optimal locations suggested by the *7*-median problem run for regular and asymmetric networks studied in this section (*A*, C_d and D_d). It can be seen that the optimal configurations exhibit only small differences from the regularly spaced solution that would be obtained with a complete network: one central location and one in the centre of each equilateral triangle (Christaller), whatever the number of edges (630 to 954). This is not the case for radial networks. Indeed, when comparing each studied solution to the latter 'average' configuration, an index of dissimilarity can easily be computed: Table 3.3 gives the number of times the optimal locations are located outside the 'optimal areas' suggested in Figure 3.19. Asymmetric networks (C_d and D_d) yield results that are similar to those suggested by the denser network (*A*). This means that network density has little influence on optimal location so long as density reduction is not too high. The rectangular network (*B*) leads to slight differences (index value equal to 2) mainly around the central point. The radio-concentric (*CR*) and radial ($R3_0$, $R6_0$) networks induce very high disparities in the optimal locations (Table 3.3): the index value is 6, meaning that only one site is located in one of the 'optimal zones' designed in Figure 3.19. The radial axes explain the differences: they attract the optimal locations.

Table 3.3 Number of supply points falling out of the 7 'optimal zones' designed in Figure 3.19

	A	B	C_1	C_2	C_3	D_1	D_2	D_3	CR	$R3_0$	$R6_0$	RH_0
Index	0	2	0	0	0	0	0	3	6	6	6	2

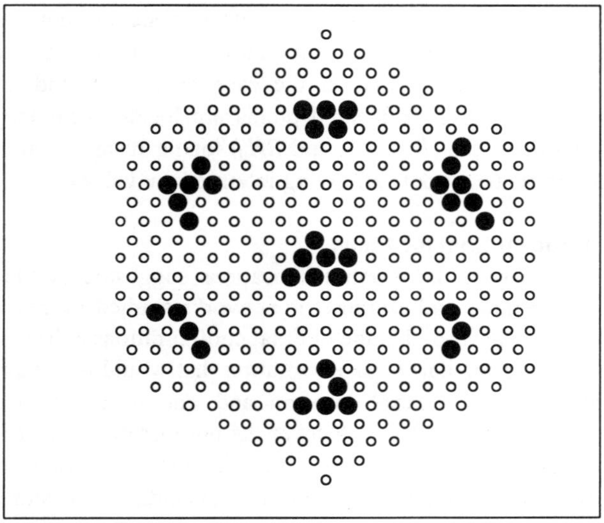

Source: *Arnold (1994)*

Figure 3.19 Optimal location of the 7 supply points for network A, B, C_d, D_d, CR, $R3_0$, $R6_0$ and RH_0

Sensitivity of the shape of the market areas

Each demand point is allocated to its closest optimal supply site. Hence each facility is associated with a zone (service or market area) including the allocated demand points and by a quantity of demand allocated to it which is the workload of the facility; this latter quantity determines the size of the facility. Both criteria (workloads of the site and shape of the service area) are used here to evaluate the *k*-median results.

Given the preceding results, the design of the optimal service areas is not surprising: regular and asymmetric networks are associated with (almost) balanced partitions of the space, while radial networks generate unbalanced partitions. Optimal locations on network *A* produce a hexagonal partitioning of the space, which recalls the Christaller's scheme. Structural variations of the network strongly affect the shape of the service areas, the presence of radial axes being of prime importance. A relationship also appears between the number of edges and the shape of the market areas, although the morphological features seem predominant. Representative graphical solutions are given in Figures 3.20 to 3.23.

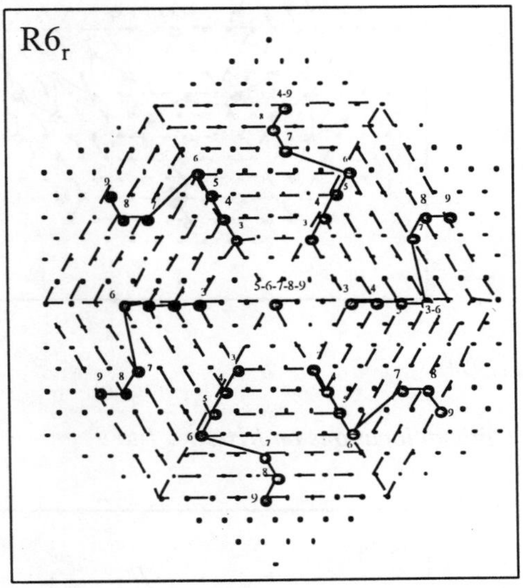

Source: Arnold (1994)

Figure 3.20 Optimal locations on R6r, r : 1 to 9

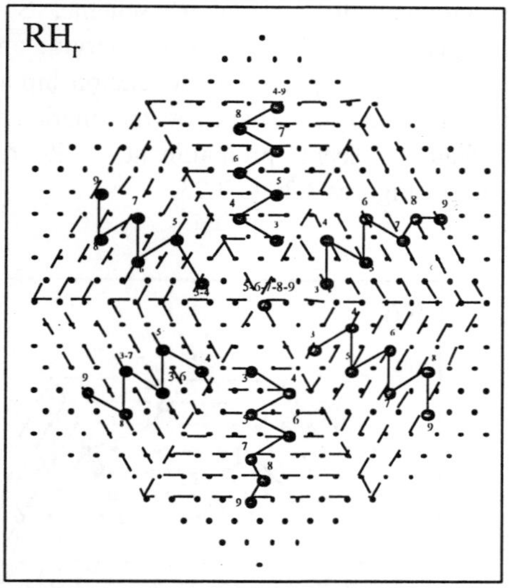

Source: *Arnold (1994)*

Figure 3.21 Optimal locations on RHr, r : 1 to 9

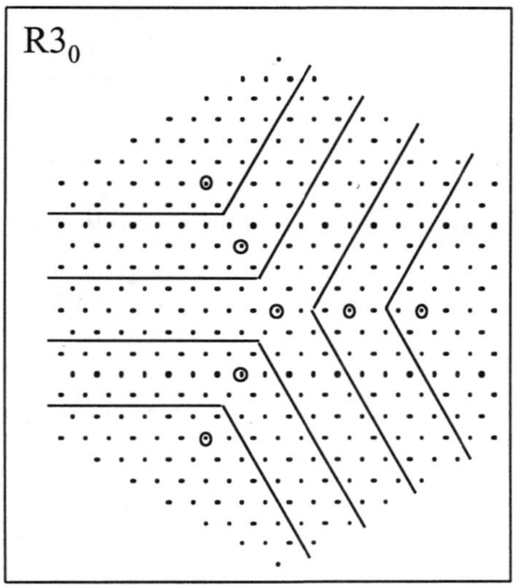

Source: Arnold (1994)

Figure 3.22 Optimal locations and allocations on $R3_0$

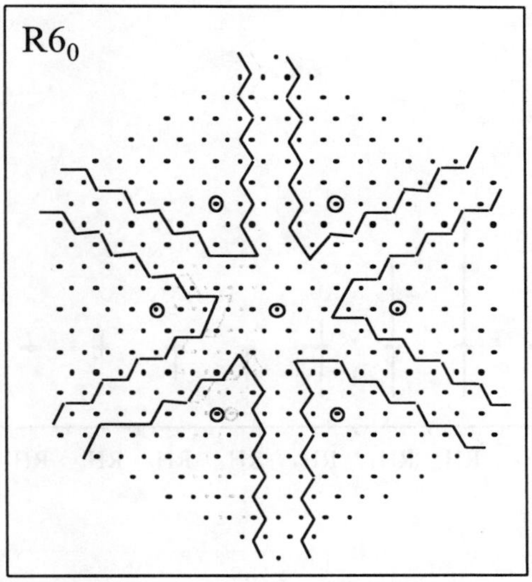

Source: Arnold (1994)

Figure 3.23 Optimal locations and allocations on $R6_0$

On radial networks, optimal locations migrate along the radial axes with *r*; sometimes (RH_r) they are located between the radials and not on them.

Allocation is here noted A_j and is estimated by the number of demand points *i* allocated to the closest supply site *j*; it measures the hierarchy of the supply sites. The value of A_j depends upon the location of the supply sites and upon the relative location of the demand points. In an isotropic environment, all sites are equal whatever *j*. Figure 3.24 shows, for each type of network, the minimum and maximum workloads of the 7 optimal sites. Small differences between minimum and maximum allocation correspond to a balance between supply sites, while large differences yield a kind of hierarchy. It shows that some networks naturally lead to a more balanced configuration (*A*, *B*, *CR*, C_d and D_d) than others ($R3_0$, $R6_0$, RH_0), illustrating again the polarisation of space induced by star-shaped networks. Density within the regular networks has little influence on the allocation disparities.

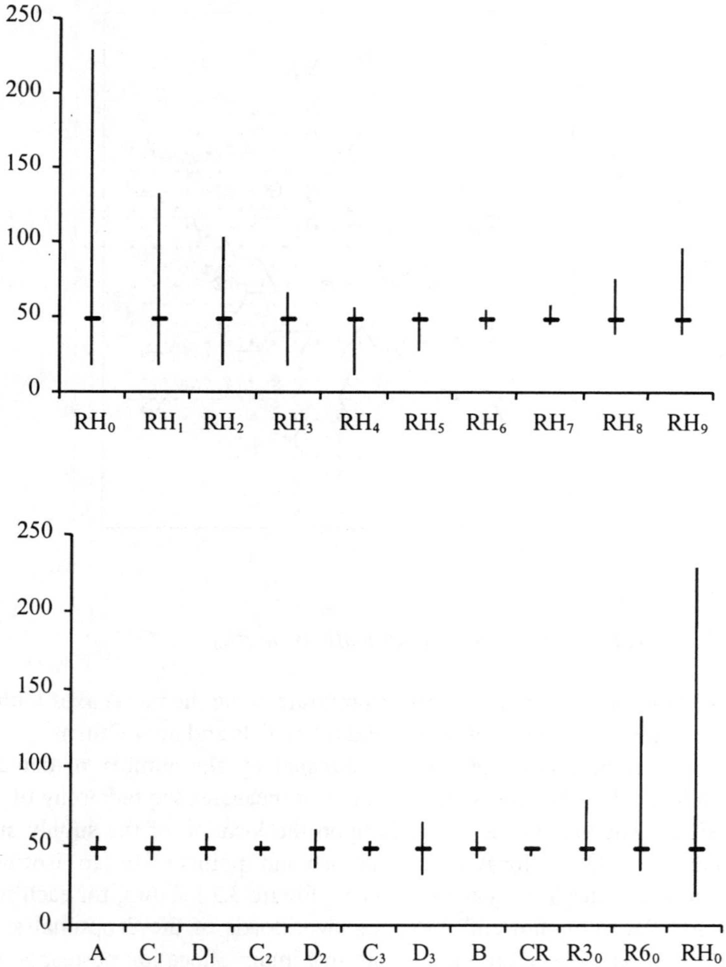

Source: Arnold (1994)

Figure 3.24 *Miminum, maximum and average workloads and type of network*

A note on the location of the ring road

The location of the peripheral road seems to be of prime importance in location results on radial networks: a good location of the peripheral leads to equity and efficiency levels comparable to more dense networks with no peripheral road. Let us go back to the preceding illustrations.

Figure 3.17 illustrates the variation of the average distance with the type of network and with the location of the ring road. For each type of network (*R3*, *R6* and *RH*), relative differences were computed between the smallest and the largest average distance; the ratio varies between 24 per cent for $R3_r$, 36 per cent for $R6_r$ and 39 per cent for RH_r. Hence a bad choice in the location of the peripheral link has important consequences in the global efficiency of the proposed solution. On the other hand, an 'optimal' choice of the location of the peripheral link enables the achievement of an efficiency level comparable to very dense regular networks; it then corresponds to a good compromise between investments in the network and efficiency of the network.

The distance frequencies also show that the most important discriminating criterion between the graphs is the location of the ring road (r): whatever the network type, locating a peripheral road near the centre ($r \leq 4$) leads to a very flat distribution of the distances, with an important range of values (many demand points are far from their closest facility). When r is greater than 4, the range is smaller, the distribution more peaked and symmetric around the mean. The location of the peripheral is thus of prime importance in the general accessibility of the supply sites.

Figure 3.24 refers to the allocation of the optimal supply sites. Whatever the network type, hierarchy between supply sites is greater when r is small, that is to say when the ring road is located near the centre of the network. The smallest differences between the maximum and the minimum workloads is reached when $r = 7$; for ring roads located further from the centre ($r > 7$) hierarchy increases again. The optimal location of the ring road in terms of allocation is $r = 7$.

Let us now consider the most important results of location-allocation models: the selected optimal supply sites (Figures 3.20 and 3.23). The position of the radial axes and the location of the peripheral road determine these locations. The migrations of the optimal locations to interaxial locations when r is large are due to border effects.

The same simulations were undertaken for other sizes of the lattice of points. The same conclusions were drawn; the only difference is the exact 'optimal' location of the peripheral link: naturally, the optimal value of r increases with the size of the studied network.

3.4.3 Comments

Let us make some additional comments on the observations made in this section. Firstly, the *k*-median model was applied to several networks designed on a fixed hexagonal lattice of points. Differences in location-allocation results for regular networks are mainly caused by variations in network density (number of edges): the denser the network, the lower the distances measured. Differences in optimal locations are slight where density changes are not too high. For a fixed density of edges, three radial networks were designed (*R3r*, *R6r* and *RHr*), generating much larger differences in the *k*-median results. The number and the location of the radial axes explain these differences: as already discussed in the literature (see Perreur and Thisse, 1974), these axes attract the optimal locations. In this section, network shape and network density were not isolated; their effects are combined. Further analyses will consider them separately.

Secondly, *k*-median results were evaluated in terms of optimal locations and allocations (market or service areas) in the geometrical space, in conformity with most real-world planning situations. The Euclidean space might not be here the right reference in order to evaluate the tested effects. Further analyses should separate the geometrical space from the topological one.

Thirdly, empirical location-allocation approaches do not leave much choice to the planner: each element of the environment is given (location of demand, volume of demand, shape of the network, and so on). Simulating network effects on theoretical networks (toy-networks) enables the control of many sources of variations and hence isolation of the tested effects. The selected environments may be seen as simplistic or unrealistic; they are, however, the only way to test the network effect, 'all other things being equal'.

The sensitivity analyses conducted in this section confirm the observations expressed in Section 3.3 for the case of a rectangular lattice of points: when a hexagonal lattice is considered, there also exists a dependency between the morphology of the network and the optimal solutions of the *k*-median problem. This relationship holds for the location of the facilities and for the distribution of distances between demand points and the service, as well as for the shape of the market areas. Although we gave evidence of a relationship between the number of edges (being chosen as an indicator of the connexity of the network) and the various characterisations of a solution, the morphological features of the network seemed predominant.

The most significant finding in this section is the contrast between the optimal solutions obtained on regular and asymmetric networks and those achieved on radial networks. The former give rise to regularly spaced facilities with similar sizes of market areas. The latter generate unbalanced supply zones, hence more disparities in accessibility to the service among clients and uneven capacities for the facilities. This, in turn, may have an impact once resources are to be distributed among facilities, leading to a hierarchy of centres. Network design has a determinant effect on spatial organisation.

3.5 CONCLUSIONS

Testing the sensitivity of the *k*-median location-allocation results to the shape of the network on several toy-networks designed on a squared and on a hexagonal lattice of points leads to the following conclusions:

1. Independently of the connectivity of the network, there is a relationship between the shape of the network and the optimal spatial organisation of the activities, whatever the studied output of the model (locations, market areas, and so on) and whatever the type of underlying lattice of points (square, hexagon).
2. Adding a peripheral link to a star-shaped network changes the location results; the location of the peripheral link is of great importance to the efficiency of the solutions. This location does not seem to be innocent: urban or regional economic growth highly depends on a good location of the peripheral ring road. It puts forward the centre–periphery problem; this result has important implications for further empirical planning work: when location is of concern, the planner has to be aware of the importance of the location of a peripheral road.
3. The shape of the market or service areas (zones) is very sensitive to the shape of the network.
4. There is a need to develop better methodologies for statistically comparing location-allocation results; up to now, no parametric test has been used and developed. This chapter – as well as the following ones – deals with descriptive comparison of measurable results.

Further simulations and considerations could be added to this set. The first is the flows induced by the network and the question of congestion: obviously, radial networks are more likely to be concerned by this issue. This question has been ignored in these sets of simulations. Another ignored question in the preceding simulations is that of capacity constraints. The second consideration deals with an uneven distribution of the populations among demand points (demand has here always been considered as equal in each point); this assumption will be alleviated in Chapter 4.

In the next chapter, new simulations will be performed in order to see if these conclusions still hold for another optimal location model (SPLP) and for other modelling conditions (non-uniform versus uniform distribution of demand; common market versus autarky).

4. Optimal locations and transportation networks: the case of a common market

4.1 INTRODUCTION

This chapter focuses on the same question as the one discussed in Chapter 3: what is the impact – if any – of the shape of the transport infrastructure on the geographical distribution of activities? In what way does the policy associated with transportation networks modify the location decisions? In order to gain more insights about the impact of transportation policy on the spatial pattern of facilities, we here consider different types of toy-networks (square lattice with different networks and different demand distributions; common markets) and a different location model. Most results developed in this chapter have been published in *Geographical Analysis* (Peeters, Thisse and Thomas, 1998). In this chapter, we study how the number and the locations of the facilities are affected by differences in the transport system; the level of activity is given. More precisely, we here focus on the simple plant location problem, the purpose of which is to find the number and locations of (public or private) facilities in order to minimise the sum of the fixed production costs associated with the setting up of facilities and of the transport costs of users to the facilities to be located.

This model is one of the prominent location problems studied in facility location analysis and very efficient algorithms have been designed to solve it (see Section 4.2.1). The simple plant location problem focuses on the trade-off between fixed production costs and transportation costs. This trade-off is at the heart of many location models (Beckmann and Thisse, 1986; Mulligan, 1984) and may be encountered in planning a system of public facilities, such as schools or recreational facilities, as well as in the design of a production–distribution–marketing strategy by a private firm (Erlenkotter, 1977). Observe also that this trade-off is central in economic geography, where it appears in an indirect way in the pioneering analyses developed by Christaller (1933) and Lösch (1940). On the one hand, the existence of scale economies at the firms' level is a critical factor for explaining the emergence of economic agglomerations. On the other hand, the need to interact among individuals and the corresponding congestion costs (defined broadly in order to include competition for land use), imply that all activities are generally not concentrated in one place. Consequently, it is fair to say that the trade-off between scale economies in production and transportation costs is critical for the geography of human activities and so regardless of the particular institutional setting in which those activities develop (see Fujita and Thisse, 1996, for a more detailed discussion). This explains why we have chosen to work in this chapter with a model focusing on that trade-off.

One may wonder why we have retained the simple plant location problem. There are at least two good reasons for that. Firstly, for any given spatial distribution of demand, the solution to this problem gives us the socially optimal configuration of facilities. Hence it is likely that our results shed light on what could be the ideal network design policy. Secondly, to the extent that the market economy yields outcomes that are (more or less) socially optimal, the solution to the simple plant location problem may be expected to be a reasonable approximation of the market solution. In this perspective, the theory of contestable markets suggests that potential competition may well discipline market economies with non-convex technologies (see Baumol, 1980, for a nice overview of this theory). For our purpose, the work by Demange and Henriet (1991) is worth mentioning since they show the existence of configurations of firms in a spatial economy such that no consumer can be made better off by changing unilaterally the strategy of a firm. In other words, our optimal configuration could well correspond to a certain type of market equilibrium.

According to the methodology adapted in the preceding chapter, we here consider various types of idealised transport networks: radial networks; circumradial networks – that is – radial networks with a peripheral road: and grid or rectangular networks. In this way, we are able to express the impact that different transportation policies may have on facility locations. For example, a radial network is likely to be associated with a centralised policy which aims at the development of a major city or conurbation (London or Paris, for instance), while a rectilinear network corresponds to a decentralised policy aiming for a more balanced spatial development (Germany, the Netherlands or the US). To be sure, most transportation networks are already built up in developed countries but there is still room for possible major modifications, like the construction of peripheral ways. Similarly, when redesigning the European commodity railway system, one should keep in mind that the choice of a particular configuration is likely to trigger a particular pattern of locations. Finally, most developing countries are still underprovided in transportation infrastructures and it is important to be able to make recommendations that agree with the long-run development objectives of these countries.

It is not clear whether or not transport infrastructure acts as a driving force in spatial economic development. Consequently, in our experiments, the possible impact that the network configuration might have on the level of activity is disregarded. In addition, since our interest is about macro-spaces, we assume that transport networks are congestion-free. The impact of the network on locations is then seized through the following variables; the observations made in autarky are identical to those obtained previously. The ones observed in the case of a common market are original.

Firstly, we retain the number of facilities. A large number of facilities may indeed be viewed as an indicator of a more dispersed configuration. Secondly, the location of facilities is also a critical variable. It is shown that radial networks have a dramatic impact on locations in that they lead to a rather small number of facilities, a single facility located at the network centre often being the only solution when fixed costs are high. By contrast, a grid (rectangular) network fosters a much more dispersed configuration with a fairly large number of facilities. These results are not terribly surprising in that they agree with casual evidence. The addition of a peripheral road reveals how the attractiveness of the centre may be weakened when a radial network already exists. The size of the ring around the centre turns out to be a critical policy parameter. This is shown by means of many numerical solutions of the simple plant location problem in the case of a uniform demand and of randomly generated local demands. Hence the addition of specific and particular links to a network may not only improve the accessibility to existing facilities but may also affect the accessibility to the facility system through a change in locations (Blum, Gercek and Viegas, 1992).

Finally, we investigate the locational implications of the opening to trade of two separated economies (a 'radial economy' and a 'grid economy') by assuming that the network of the common market is now given by merging the two national networks. This is an important question from the policy standpoint since the creation of a single market involving economies where different national transportation policies have been followed (for example, France and Germany or the United States and Mexico) is occurring both in North America (NAFTA) and in Europe (the EU). There is a growing literature devoted to the impact of globalisation on the geographical distribution of activities (see the recent survey by Ottaviano and Puga, 1998). However, this literature fails to deal explicitly with the shape of transportation infrastructure. This is what we investigate here in the hope of providing useful insights. When the long-run locational adjustments have taken place, it is seen that the regional grid economy still involves several facilities whereas the regional radial economy is characterised by a much smaller number of facilities. The border between the two regional structures turns out to be critical in that it attracts facilities on the side of the grid economy. These results shed new light on the possible geography that could emerge within a common market. They further suggest that history matters for locational patterns as far as transport networks are concerned: the formation of a common market is not likely to have a dramatic impact on the regional structures of locations, even though the absolute level of activities within each region may well be significantly affected. This conclusion is strengthened by our complementary study of the impact of the network density. Combining two economies endowed respectively with a dense and a sparse grid network (somewhat like the case of West Germany and East Germany), we observe that the shape of the network seems to have a stronger impact than its density.

4.2 MODEL AND EXPERIMENTS

4.2.1 The SPLP in Operations Research and Economic Geography

A great deal of attention has been paid in the 1970s and the 1980s by operations researchers and management scientists to the modelling of locational decisions. There now exists a well-developed body of literature which is of great interest to economists, geographers and regional scientists (Francis and Mirchandani, 1990 and Drezner, 1995 contain several surveys devoted to various aspects of locational decision making).

A particular model, known as the 'simple plant location problem' (SPLP), has emerged as the main prototype. This model captures the trade-off emphasised in the work of Christaller and Lösch: given some requirements for a composite good distributed over space, the purpose is to determine the number and locations of facilities so as to minimise the sum of production and transportation costs. A thorough survey presenting the structural properties of the problem can be found in Cornuejols, Nemhauser and Wolsey (1990).

On the demand side, social needs are expressed by some fixed requirement for a composite good. Requirements are distributed over a finite number of points $j = 1, ..., m$ and the requirement for the composite good in site j is denoted by δ_j. On the supply side, facilities can be placed at a finite number of potential locations $i = 1, ..., n$ while production involves scale economies. The set-up cost F_i and marginal cost c_i are constant; hence the production costs of a facility at i with output q_i are given by $F_i + c_i q_i$. Different fixed costs (F_i) may account for differences in fixed factor endowments, while the marginal costs ($c_i \times q_i$) may reflect particularities in local competition for variable production factors. Finally, the cost of shipping one unit of the composite good from site i to site j is a constant. Clearly, the matrix (t_{ij}) of transportation costs is general enough to allow for different shapes of the transportation network and various access conditions to local markets. Note that a rise in fixed production costs is formally equivalent to a fall in transportation costs; hence studying the impact of the F_i on the locational pattern amounts to studying the impact of the t_{ij}.

Formally, the SPLP is defined as follows:

$$\text{Min} \sum_i \sum_j (c_i + t_{ij}) \delta_j x_{ij} + \sum_i F_i y_i \tag{4.1}$$

subject to

$$0 \le x_{ij} \le y_i, \forall i, j \tag{4.2}$$

$$\sum_i x_{ij} = 1, \forall j \tag{4.3}$$

$$y_i \in \{0,1\}, \forall i \tag{4.4}$$

where x_{ij} stands for the (non-negative) fraction of the demand at j supplied by a facility at i and y_j a 0 – 1 variable which equals 1 when a facility is set up at j and 0 otherwise. The first set of constraints implies that no demand can be supplied from a site where no facility has been built. The second set of constraints means that the total requirement in each i must be met. The program used is the one developed by Hanjoul et al. (1990).

4.2.2 Lattice, Networks and Demand Structures

Computational experiments are once again conducted on idealised networks. Choosing a theoretical lattice rather than a real-world layout enables us to better isolate the tested problems from many other sources of variation and to control as much as possible for spatial layout. A hypothetical settlement allows us to conduct tests without the additional complexity introduced by empirical modelling; it is completely free from any contamination related to data collection, coding or other errors and therefore should lead to a greater generality of the conclusions.

As discussed above, we consider simulations on a discrete transportation space. The set of points chosen among the possible spatial configurations is a 17 × 17 squared lattice of 289 points where each point j has the same spatial environment as every other, at the exception of the border points. Each point j of the lattice is simultaneously a demand point and a potential location for a facility. Each j is characterised by its coordinates (x_j, y_j) and is linked by an edge to its closest neighbours. Each link has a length of 100 for horizontal or vertical edges and 141 for diagonal edges. Note that some experiments have been conducted on larger networks; results do not change.

A quantity δ_j is associated to each point j of the lattice. It stands for the quantity demanded of the composite good in that point. In a first set of simulations (see Section 4.3.1), this quantity is supposed to be invariant with j, that is, the distribution of demand is uniform, with $\delta_j = 3$ for all j (where the value 3 has been chosen in order to make the comparison with the stochastic demand case easier). In a second set of simulations (see Section 4.3.2), demand varies randomly across locations. Specifically, the support of δ_j is given by the integers between 1 and 8, the total demand ($\sum_j \delta_j$) being the same in both sets of simulations. We allow for substantial variations in the spatial pattern of local demands in order to gain more robustness in our results.

The number of edges and their spatial organisation define the transportation network. Three families of transportation networks are designed on the squared lattice of points. Our choice is motivated by the fact that the selected networks approximate real-world networks fairly well. More precisely, we consider grid (*Rect*) and (circum-)radial networks ($Circ_r$, $r \geq 0$). In network *Rect*, only the horizontal and vertical edges are considered as transportation links, thus corresponding to a grid network (see the right-hand size of Figure 4.1). Radial networks have fewer links than regular networks. $Circ_0$ represents a radial network without any peripheral link (see the left-hand size of Figure 4.1). It is well known that this type of network tends to favour the centre for accommodating a facility location (Perreur and Thisse, 1974). When adding a belt to a radial network, one obtains a circumradial network ($Circ_r$, $r > 0$). In this type of network, accessibility is improved because there are more edges. Perreur and Thisse (1974) then show that optimal facility locations may also occur on the peripheral road, that is, at the intersections with the radial axes. In the preceding chapter, we have shown that the position of the peripheral road is of prime importance for the choice of facility locations in the k-median problem. The same result is expected to hold for the simple plant location problem. In the present chapter, r is an index that stands for the position of the ring road. When r is equal to 1 the ring road is close to the network centre; when r is equal to 8, the ring road is located at the outer limit of the lattice.

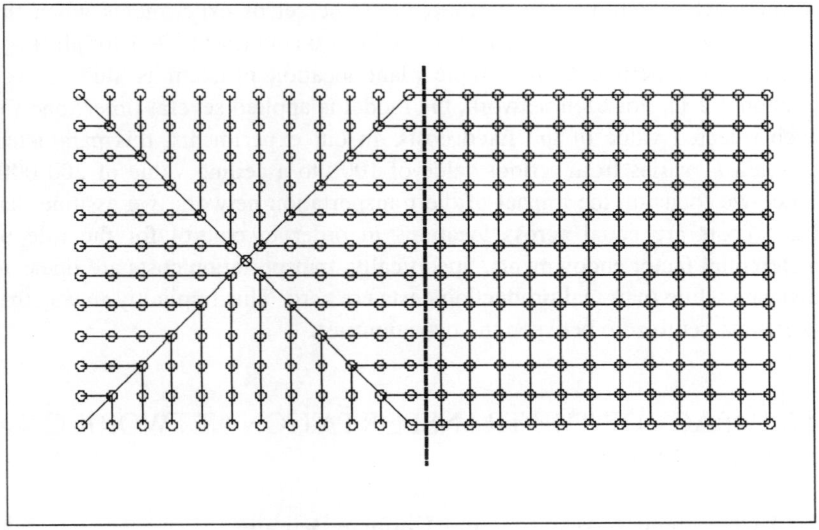

Figure 4.1 Map of the mixed network (common market)

Last, we merge two squared networks in order to see how far the optimal locations in a common market depend upon the underlying regional networks. The mixed network is made of two squared lattices (Figure 4.1). Simulations have been performed on several sizes of rectangles and show that the size of the network does not affect our results. So we here refer to a rectangle of 289 points, made of two 12 × 12 squared lattices. On the left squared lattice, a radial network is designed and a central point is added (145 points); on the right one, a grid type network is designed on the 144 remaining points. Finally, the two squared lattices are bound together by 12 horizontal links. In this set of experiments, demand is supposed to be the same across locations.

We have also investigated the question of dense versus sparse networks by merging two grid networks, one of which is described by a 15 × 15 squared lattice and the other by an $l \times l$ squared lattice, where l stands for the number of horizontal or vertical links and takes respectively values from 2 to 8, while the length of the corresponding edges is adjusted for the covered area to be the same. In so doing, we want to control for the country size effect. Firstly, we assume that the global level of activity is the same in both economies in order to isolate the sole impact of the density of the network. This implies that local demand in the sparse network is higher than in the dense network (for example, when $l = 5$, $\delta_j = 9$ in the sparse network whereas $\delta_j = 1$ in the dense network). Secondly, since a sparse network is often associated with a lower level of activity, we perform the same set of experiments when the local demand is the same everywhere in the two countries ($\delta_j = 1$ for all j).

In each experiment, the simple plant location problem is studied (see Section 4.2.1). For each network, the model is applied several times, one for each selected value of the fixed costs. In our experiments, this parameter, denoted F, varies from a floor value of 1000 to a ceiling value of 200 000. Since we focus on the impact of the transportation network, we assume that fixed costs are equal across locations in order to control for the role of differential factor endowments. Specifically, transportation costs are linear in distance while marginal production costs are zero. This implicitly means that tariffs are set to zero between the two countries.

4.3 IMPACT OF THE TRANSPORTATION NETWORK ON LOCATIONS

4.3.1 Comparing Networks under Uniform Demand

We here consider a squared lattice of 289 vertices over which demand is supposed to be uniformly distributed; in this way, we concentrate on the sole geographical implications of the network configuration. The four networks described above are studied.

Number of facilities and level of fixed costs

Figure 4.2 gives the optimal number n^* of facilities for F varying from 5000 to 200 000 and for the four networks. As expected, the optimal configuration of facilities contains less and less facilities when F rises: when the fixed costs are large, activities tend to concentrate fewer places. In the radial network case, we see that the decrease in n^* is very sharp; a single facility solution is obtained from a fairly low value of the fixed cost, thus suggesting that the centre of a radial network is the source of a strong agglomeration force. On the contrary, in the case of the grid network, the optimal configuration involving a unique facility is obtained with a value of F which is three times as large as in the radial case, confirming the intuition that such a network yields more dispersion in human activities. For circumradial networks, we obtain intermediate solutions. This means that the construction of a peripheral road is indeed an instrument that can be used by spatial planners in the aim of fostering a more scattered distribution of human activities. Observe that the analysis of the values taken by both the transportation costs and installation costs confirms the results presented in Figure 4.2 in that transportation costs rise with the level of fixed costs through a fall in the number of facilities. Though all these observations depend on the fact that F is the same across locations, no major change is expected to arise when variations in fixed costs are not too large.

For radial and circumradial networks, the decrease in F occurs by steps; this is due to the underlying geometry.

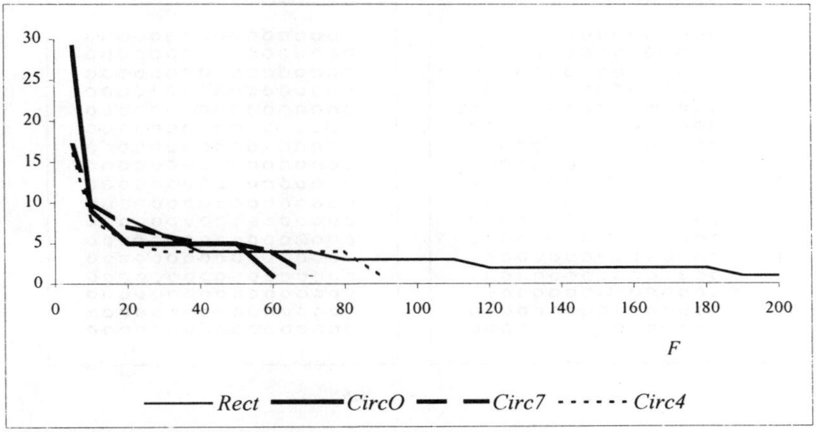

Figure 4.2 Variation of the number of facilities with F

Optimal locations

We now turn to the characterisation of the optimal locations. We retain the value F = 80 000 which leads to typical results in our many experiments. Starting with a radial network, we observe immediately that a single facility is set up at the centre (Figure 4.3). This is reinforced by the fact that there is always a facility at the centre for all admissible values of F. On the other hand, a grid network leads to the construction of three facilities which are (more or less) evenly spaced (Figure 4.3).

Interestingly, if a peripheral road is installed at some intermediate distance from the centre ($r = 4$), the optimal configuration then involves four facilities, all located at the crossing of the radial roads and of the ring road (Figure 4.3). This shows that the attractiveness of the centre may completely vanish when a peripheral road is built at a well-chosen distance from the centre. Again, such an observation is confirmed for other admissible values of F. It is worth noting that the choice of the radius r is crucial for this result: in our simulation set where r varies from 1 to 8, we can show that the centre no longer accommodates a facility when $2 \leq r \leq 4$, but it is included in the optimal configuration for the other values of r. For low values of r, the ring road has almost no impact on the optimal pattern of locations because the nodes it generates are too close to the centre. On the other hand, large values of r are such that these nodes are now situated at the outskirts of the area so that they have only small hinterlands to supply, a fact that strongly reduces their attractiveness.

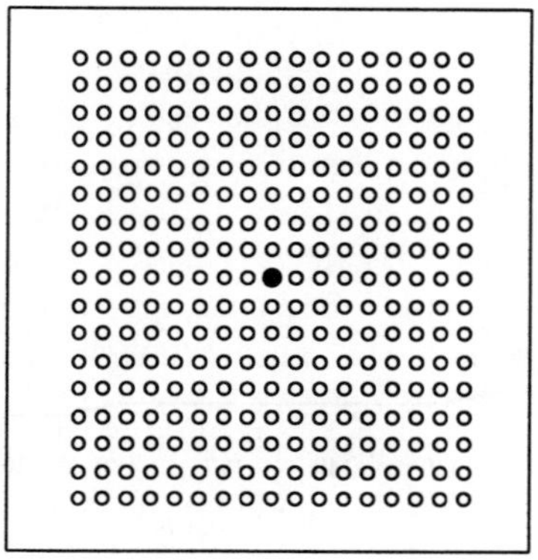

$Circ_0$

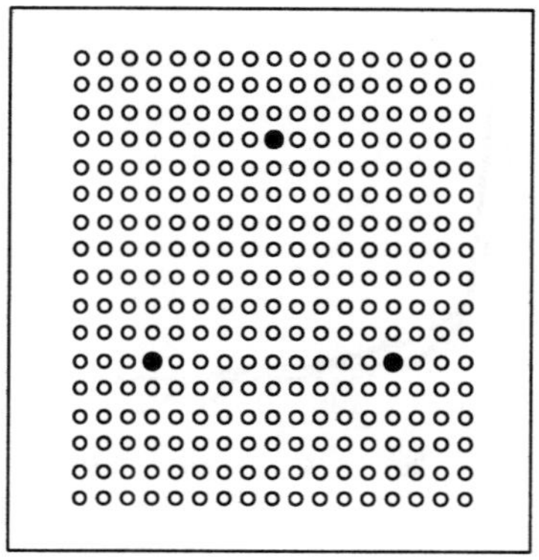

Rect

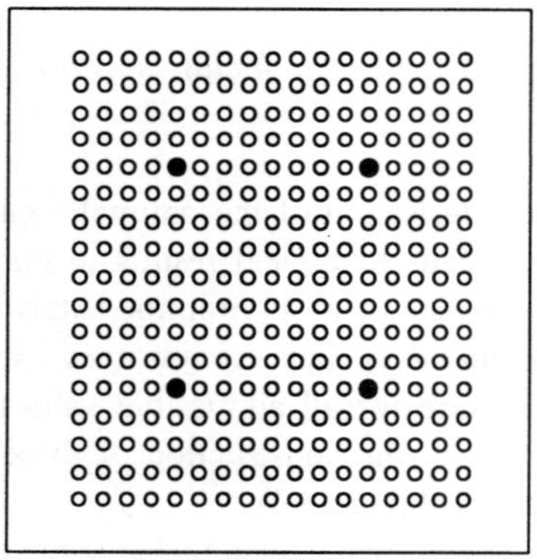

$Circ_4$

Figure 4.3 Optimal locations when $F = 80\ 000$ for a radial ($Circ_0$), a grid (Rect) and a circumradial network ($Circ_4$)

4.3.2 Comparing Networks when Demand is not Uniform

Let us now alleviate the assumption of the uniform spatial distribution of demand. Numerous distributions of demand can be considered on the studied squared lattice of points. We here refer to only two distributions: a random distribution and a pyramidal distribution. First, we have generated several sets of random values of demand such that the total demand associated with all points ($\Sigma_j \delta_j$) is the same as the one previously used (uniform). We here refer to only one of these generated distributions. In the case of the pyramidal distribution, the centre of the lattice is characterised by a very high value of demand, the periphery by very low values; the decrease in demand is regular with distance to the centre. The total demand associated with all points ($\Sigma_j \delta_j$) is once again the same as in the random distribution.

In order to gain more insights into the robustness of the results displayed in the foregoing section, we first assume that the requirements are generated randomly. Roughly speaking, the results are quite similar to those obtained with a uniform demand. It is seen that, regardless of the network configuration, the number of facilities is almost the same under the two demand structures (Figure 4.4). The same holds for most optimal locations. Consequently, the shape of the transportation network seems to have a stronger impact on the optimal facility locations than the spatial distribution of demand. This is rather unexpected in view of the amount of work devoted to the impact of demand in one-dimensional location models and given the large variety in local demands we have allowed for in our experiments.
When considering the pyramidal distribution of demand, stronger differences are observed. Figure 4.5 illustrates the decrease in n^* with F for the radial network with a peripheral in $r = 5$ and the three types of demand distributions. As mentioned in the preceding paragraphs, uniform and random distributions lead to similar location results. Pyramidal distributions are slightly different: the decrease in n^* with F is sharper leading to one facility for a smaller value of F; in this case, the attraction force of the centre is reinforced by the superposition of two similar spatial organisations: the centripetal transportation network and the centre–periphery distribution of demand. Note that for values of F equal to 20 000, five facilities are located; they are exactly the same, whatever the demand structure: one central location and one location at the crossing of the radial axes with the ring road. In all other cases, locations are slightly different.

Given the results obtained in this section, the simulations performed in the next section will be limited to the uniform distribution of demand.

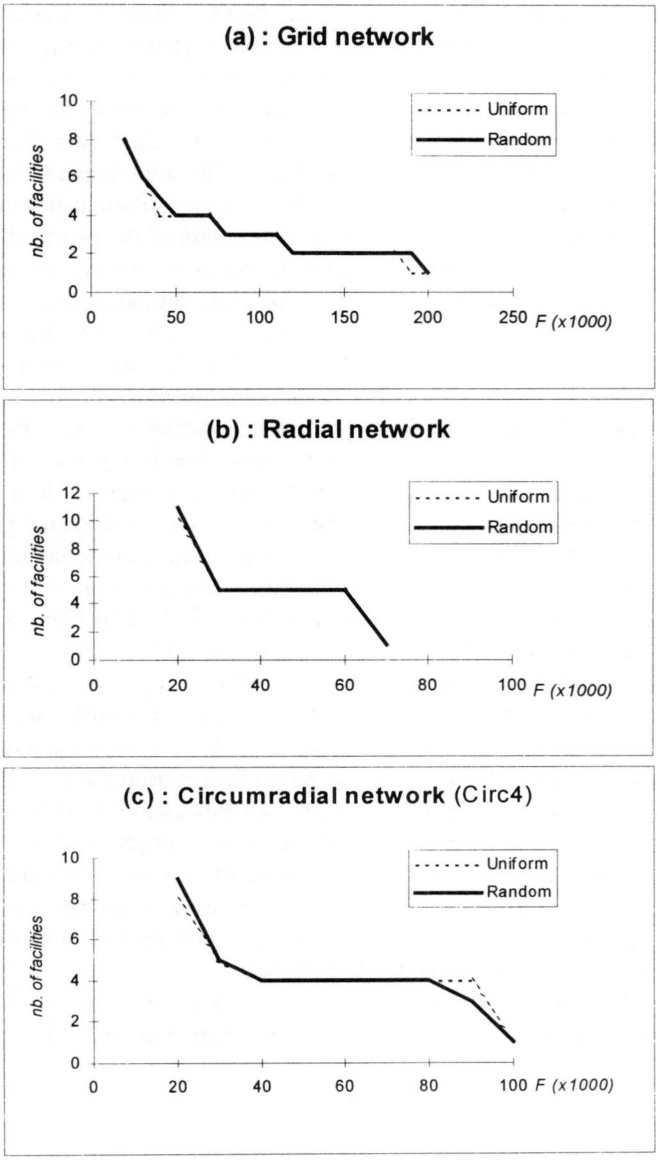

Figure 4.4 Variation of the number of facilities with F, for 3 networks and 2 types of demand distribution

4.3.3 Merging Different Networks in a Common Market

In this section, we want to evaluate the impact of the network configuration when two isolated economies characterised by a different network are integrated or linked. For the reasons discussed in the introduction of this chapter, we focus on radial and grid networks that are adjacent. The total size of the integrated economy is the same as that of a single economy (289 points and uniform demand). We assume that both economies have the same size in order to control for the size effect, and that the number of 'gates' (see Chapter 5) between the two economies is equal to the common number of horizontal links in both networks.

In Figure 4.6, we give the optimal number n^* of facilities in the integrated economy as well as in each separate economy. For small or large values of the fixed cost F, the total number of facilities is about the same. In other words, with low– and high– scale economies, the opening of the economies is not likely to have a strong impact on the locational pattern (assuming of course that the spatial distribution of demand remains the same). On the other hand, for intermediate values of F, the integrated economy typically involves a number of facilities smaller than the total number of facilities operating in the separate economies. In this case, market integration seems to yield more geographic agglomeration.

Turning to the optimal locations, two cases are considered: Figure 4.7 describes the optimal locations when both economies are integrated, while Figure 4.8 corresponds to the case where the two economies are separated. For presentation reasons, these figures are at the end of the chapter. We may summarise our main results as follows. Firstly, the centre of the radial subnetwork attracts a facility in both the integrated economy and the separate radial economy. Secondly, the grid regional economy often admits a larger number of facilities than the radial regional economy. Hence even when a grid economy is integrated with a radial one, it retains most of its properties regarding the dispersion of human activities.

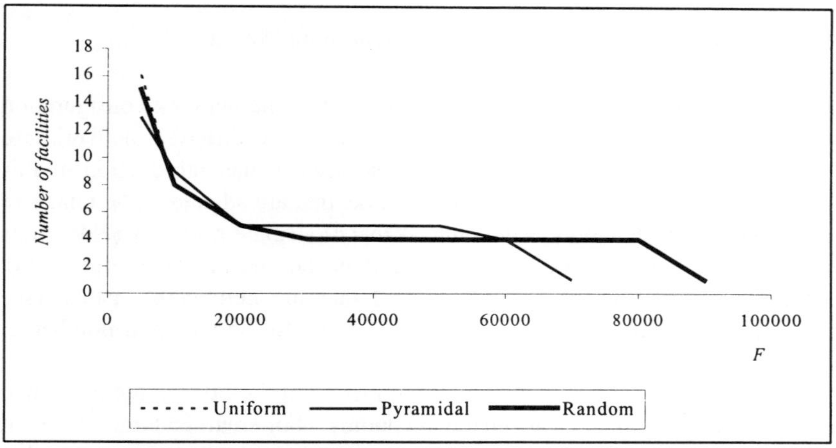

Figure 4.5 Variation of the number of facilities with F, for a circumradial network (r=5) and 3 types of distributions of demand

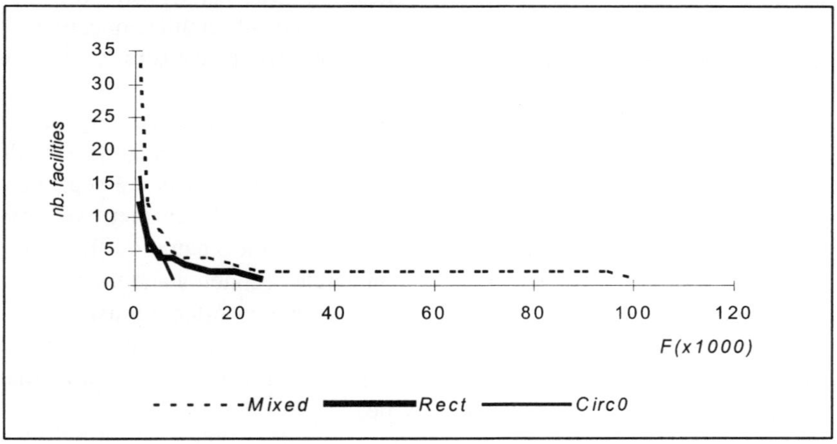

Note : Rect: 144 points, *Circ*$_0$*:145, Mixed*: 289 points

Figure 4.6 Variation of the optimal number of facilities with F and the type of network

Thirdly, for values of F not too large, Figure 4.7 ($F = 5000$ and 7500) shows that some facilities set inside the grid regional economy are established near the border between the two economies (think of Strasbourg or of some segments of the American–Mexican border). Thus though the formation of a single market has an impact on locations, these are primarily to be found in the region endowed with a network inducing the dispersion of activities. Hence the history of national transportation policies should matter for the geography of the integrated economy. We will return to this point in the next section. Lastly, recall that tariffs are supposed to be zero in the custom union since transportation costs are unaffected by the crossing of the border. The existence of positive tariffs would lead to configurations closer to those obtained in each separated economy (Figure 4.8).

4.3.4 Merging Similar Networks with Different Densities

We now consider the case of a 15 × 15 grid network (left–hand size) together with an $l \times l$ grid network (right–hand size) in the hope of evaluating the impact of network density, by varying l from 2 to 8 (Section 4.2.2). Figures 4.9, 4.10 and 4.11 refer to the case where total demand on the right–hand side (RHS) is equal to the total demand on the left–hand side (LHS). Figures 4.12 and 4.13 refer to the case where demand is set equal to 1 in all j's. Figure 4.12 refers to the results of merging a 15 × 15 lattice with an 8 × 8 lattice of points and Figure 4.13 with a 3 × 3 lattice.

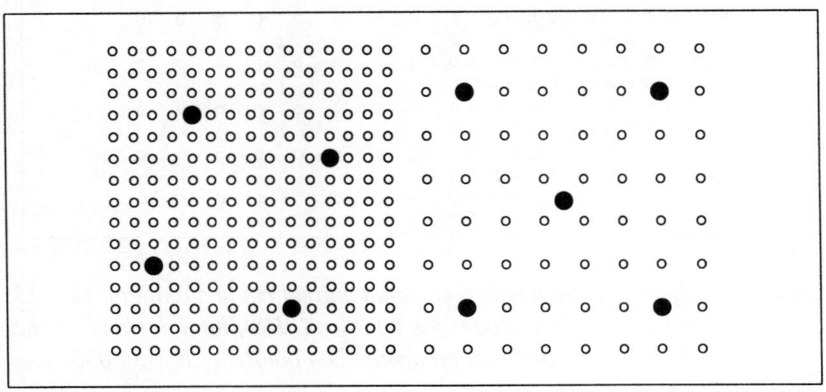

Figure 4.9 Optimal locations in a common market made out of one 15 × 15 lattice (LHS) and one 8 × 8 lattice (RHS), total demand on the LHS being equal to total demand on the RHS. F = 10 000

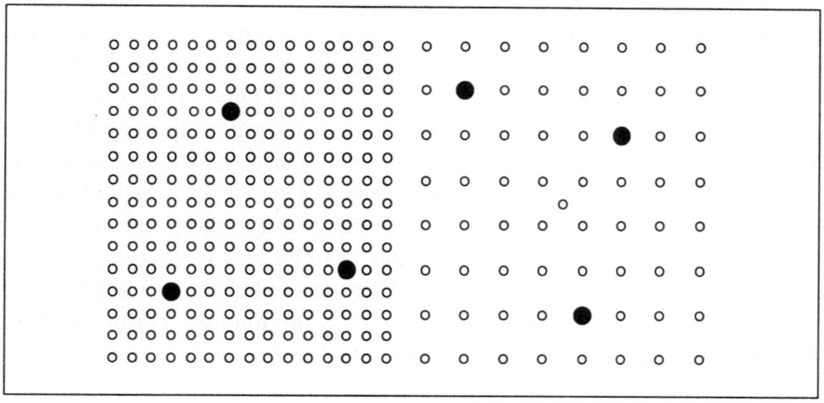

Figure 4.10 Optimal locations in a common market made out of one 15 × 15 lattice (LHS) and one 8 × 8 lattice (RHS), total demand on the LHS being equal to total demand on the RHS. F = 20 000

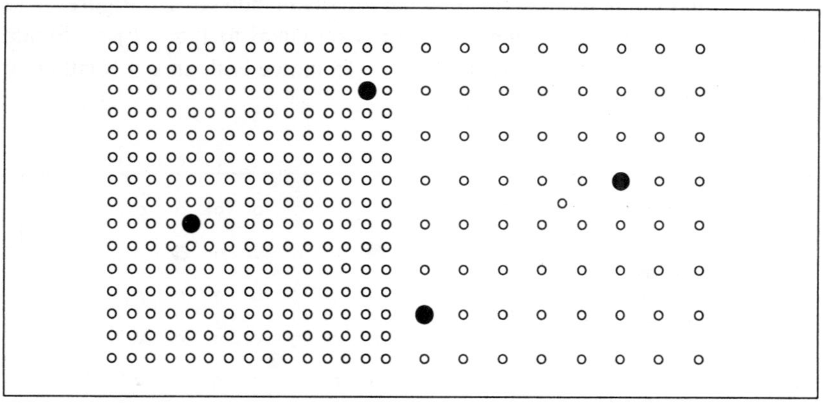

Figure 4.11 Optimal locations in a common market made out of one 15 × 15 lattice (LHS) and one 8 × 8 lattice (RHS), total demand on the LHS being equal to total demand on the RHS. F =30 000

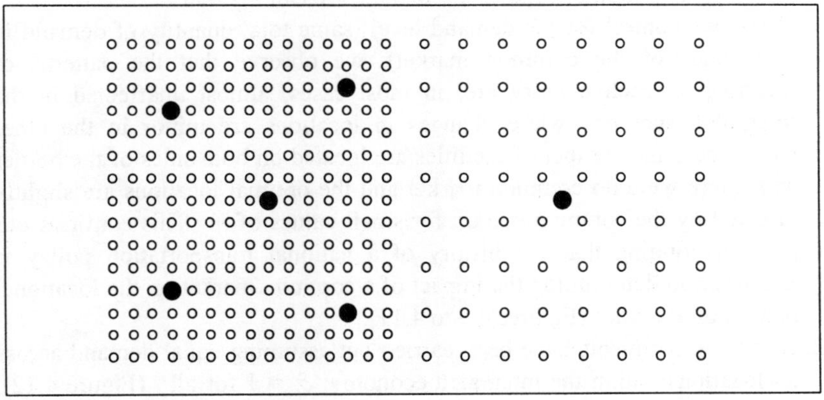

Figure 4.12 Optimal locations in a common market (15 × 15 LHS + 8 × 8 RHS), demand being equal to 1 ∀ j. F = 10 000

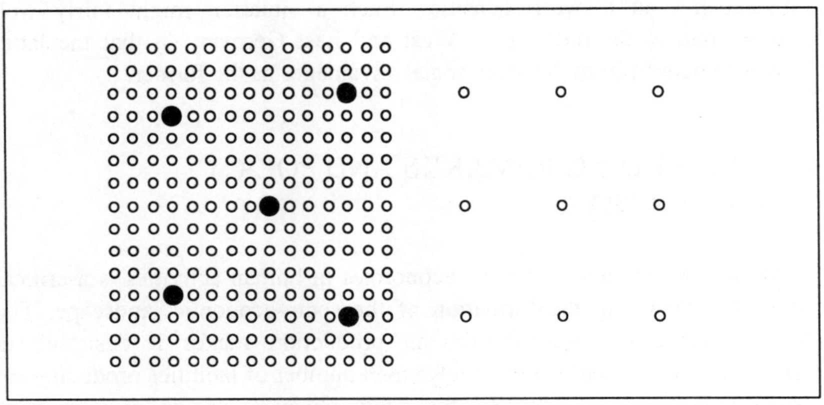

Figure 4.13 Optimal locations in a common market (15 × 15 LHS + 3 × 3 RHS), demand being equal to 1 ∀ j. F = 10 000

From the observation of Figures 4.9 to 4.13, we can say that:

1. When we control for the demand level (same total quantity of demand in both sides of the common market), we observe that the patterns of locations in each country are, in most cases, almost unaffected in the integrated economy while changes in locations are minor in the other cases: the same number of facilities are located on both sides of the border as if there were no common market and the optimal locations are slightly affected by the common market for small values of F. This confirms our previous finding that the history of a national transportation policy is important in determining the impact of a common market on the locational pattern of activities (Figures 4.9 to 4.11).
2. Similar experiments have been carried out assuming equal demand across all locations within the integrated economy: $\delta_j = 1$ for all j (Figure 4.12). Unless fixed costs are very small, we observe some predatory effects at the expense of the sparse network: the dense network economy dominates the sparse network economy whatever its size. The latter experiences a substantial decrease in the number of facilities; in the limit, the corresponding country empties out (Figure 4.13). Clearly, this dramatic effect is due to the fall in the number of demand points and not much to the discrepancy in network densities. Such a situation might fairly well correspond to the merging of West and East Germany in that the latter tends to benefit from the same social advantages as the former.

4.4 CONCLUDING REMARKS AND POLICY IMPLICATIONS

Recall that the existence of scale economies in human activities is a critical factor for explaining the formation of the socio-economic landscape. The mere existence of indivisibilities in production makes it desirable to concentrate production in a relatively small number of facilities producing for dispersed consumers. However, the geographical extension of interaction areas implies the existence of positive transportation costs (defined broadly in order to include all the impediments to trade), implying that the entire production is generally not concentrated in one place. Therefore, there is a fundamental trade-off between scale economies and transportation costs in a spatial economy, whence our interest in the simple plant location problem. Since the beginning of the Industrial Revolution, the increasing spatial concentration of activities has been caused by the fall in transportation costs and the rise in fixed production costs (Bairoch, 1985), two facts captured by the SPLP.

It is our belief that obtaining general results regarding the impact of the shape of the network on the location of facilities is highly problematic because of the combinatorial nature of the problem, even in the case of toy-networks. However, we also believe that the many experiments we have made provide us with particular, but meaningful, results. Several comments are now in order. Firstly, the outcome of the spatial trade-off is strongly affected by the shape of the transportation network, thus confirming the robustness of the localisation theorems obtained in facility location analysis (Beckmann and Thisse, 1986; Labbé, Peeters and Thisse, 1995). More precisely, we have seen that radial networks lead to more concentrated patterns of production, the centre playing the role of a powerful attractor. In contrast, a grid network leads to a more dispersed pattern with several locales accommodating facilities. Though such results probably belong to the folk wisdom of human geography, it seems to us that our many experiments give them more robustness.

However, radial economies are not necessarily stuck with concentrated geographical patterns. The construction of a peripheral road around the centre of a radial network can boost a major deconcentration of human activities towards the nodes situated at the crossings between the radial roads and the ring road. If such a policy is to be implemented, the choice of the ring radius turns out to be a critical variable. Too small a radius leads to a circumradial network that does not depart enough from the original configuration to generate dispersion, while a large radius yield nodes with too–small hinterlands for them to become attractive. Clearly, even when the radius of the peripheral road is well chosen, it would be illusive to expect the outward pull discussed above to generate its effects in the short or medium run. The possible impact of such a policy on locations will become visible only in the (very) long run.

This conclusion must be qualified in view of some recent contributions in spatial economics. As discussed by Arthur (1990) and Krugman (1991), human activities may be locked-in at some particular places for reasons that have nothing to do with the transportation network (examples are provided by the 'industrial belt' in the US or the 'golden triangle' in Europe). Indeed, it seems that modern economies are more and more characterised by a 'putty-clay geography' in which there is a priori a great deal of flexibility in the choice of locations but a strong rigidity in spatial structures once the process of agglomeration has started. The forces generating lock-in effects are based on the spatial interdependence between consumers and producers. They are not present in the SPLP where the spatial distribution of demand is fixed. Such forces should normally reduce the impact that the construction of a peripheral road might actually have on locations, even in the very long run. Yet, it remains true that the main nodes of the transportation network are likely to be the focal points of an economic geography in which human activities become more and more footloose (a spectacular example of such a phenomenon is provided by Chicago; see Cronon, 1991). Stated differently, major nodal points chosen in the past as locations of facilities will probably be the most advantageous places in the future, because most of them have experienced the accumulation of a wide array of comparative advantages that will permit them to remain attractive. Note also that the spatial structure of demand is supposed to be fixed in our experiments. It seems reasonable to believe that some individuals will choose to relocate in order to be closer to the facility locations. This opens the door to the mobility of households, which may vary across countries or regions.

Secondly, the role of transportation infrastructure in an integrated economy has been a very much neglected topic. To the extent that the construction of transportation infrastructures can be regarded as irrevocable, networks are hostages to the past and keep influencing the decisions made in the new economic and political environment. This legacy is heavy: within the limits of the regional economy, a network has about the same impact on the spatial organisation of activities as in the case of autarky. Even though the level of activity is likely to be affected by the integration of various economies, it is less clear that the geography of human activities within each one of them is to be drastically modified (Krugman, 1991). On the contrary, the developments above suggest that the geographical organisation of economic and political activities might well remain approximately the same within each original territory. Accordingly, one must be careful in drawing clear-cut implications about the impact of economic integration on the spatial distribution of human activities. Indeed, transportation networks, especially radial networks, are to be viewed as a strong force of inertia in the location of human activities, that could supplement those discussed above.

Lastly, network configurations have long-run implications for the spatial organisation of human activities. This is a strong message for developing countries that invite us to reconsider standard policies. Radial networks have often been adopted in such countries (Alonso, 1969) and there are good reasons for such a choice. Firstly, radial networks are organised around the main metropolitan area of the corresponding country, that is, around a place with much need. Secondly, they require less resources to be built than grid networks, but we did not consider the total cost of construction of a network. However, what seems to be a sensible decision from the short– or medium–run point of view may no longer be reasonable in the long run because the resulting balance of activities in space is very uneven. This suggests a gradual policy in which irrevocable decisions regarding the shape of the transportation network are delayed in the hope that more resources will become available for the construction of a grid–like network covering the whole territory under consideration.

A second warning regarding policy is in order. Krugman and Elizondo (1996) have studied the impact of trade policy on urban concentration in developing countries. They show that there is a surprising linkage between trade policy and urban development: free trade would foster urban decentralisation while restrictive trade practices would induce more spatial agglomeration. To the extent that local production is more and more directed to the world market, access to the main domestic market becomes less crucial than in autarky, thus reducing the lure of the primatial city. To obtain their result, Krugman and Elizondo assume that the developing country is formed by two regions with no spatial extension and no transportation networks. Here also, our results suggest that strong policy recommendations based on such simple models are to be reconsidered within a more general framework accounting for transportation infrastructure.

Our analysis is incomplete in several respects. Among the possible extensions that are worth studying, we would like to mention the following. Firstly, using general networks, one may study the possible impact of the construction of additional links as well as the increase of the capacity and quality of existing links. This is especially relevant in the case of the merging of several economies since national transport policies typically neglect the global impact of international connections. Secondly, we could improve upon our framework by assuming that national networks are connected through specific gates. Clearly, an uneven distribution of gates should have a significant impact on locations. Thirdly, we restrict ourselves to a small number of toy-networks but our approach could be extended to other typical configurations. Fourthly, when the focus is more on micro-spaces such as urban areas, one should account for the presence of congestion along some edges when assessing the impact of new links.

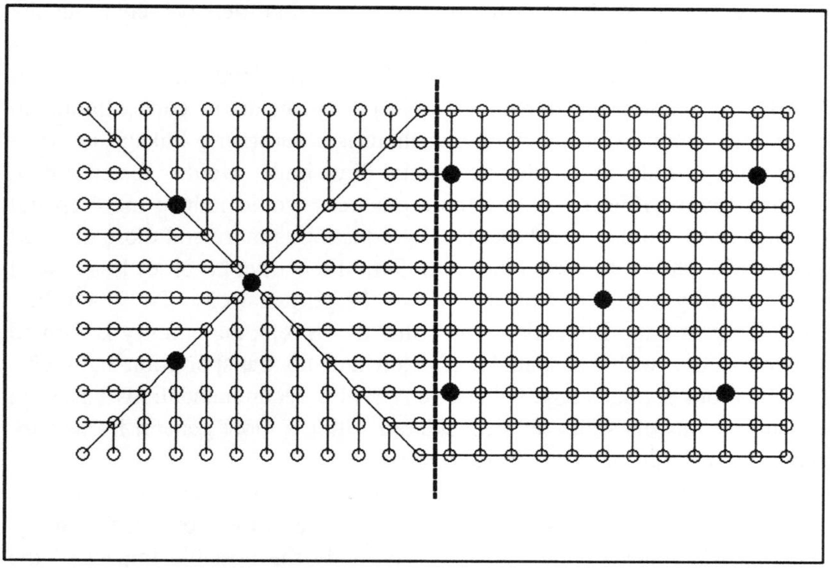

$F = 5000$

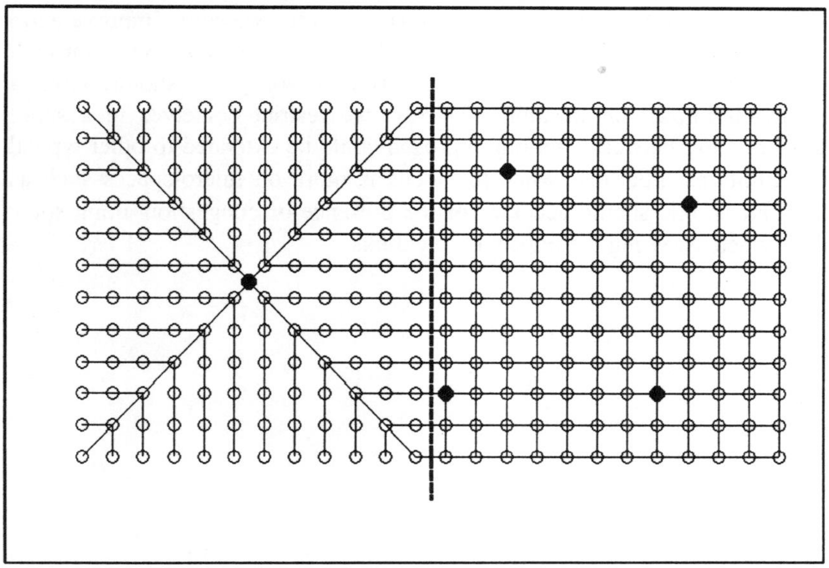

$F = 7500$

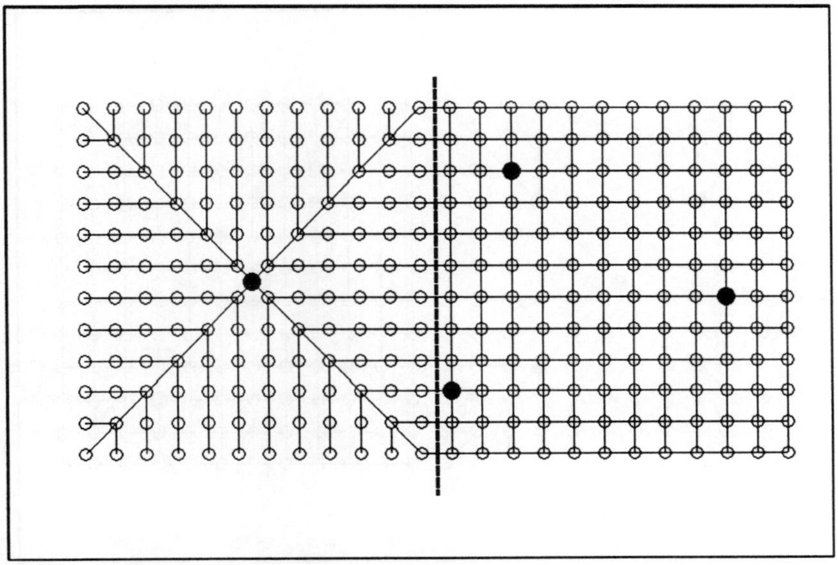

$F = 10\ 000$

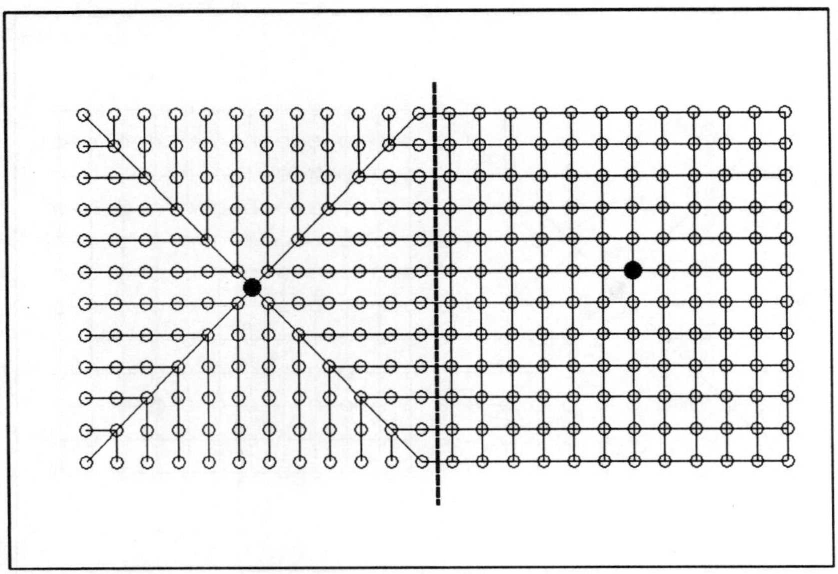

$F = 30\ 000$

Figure 4.7 Optimal locations in a common market for different values of F

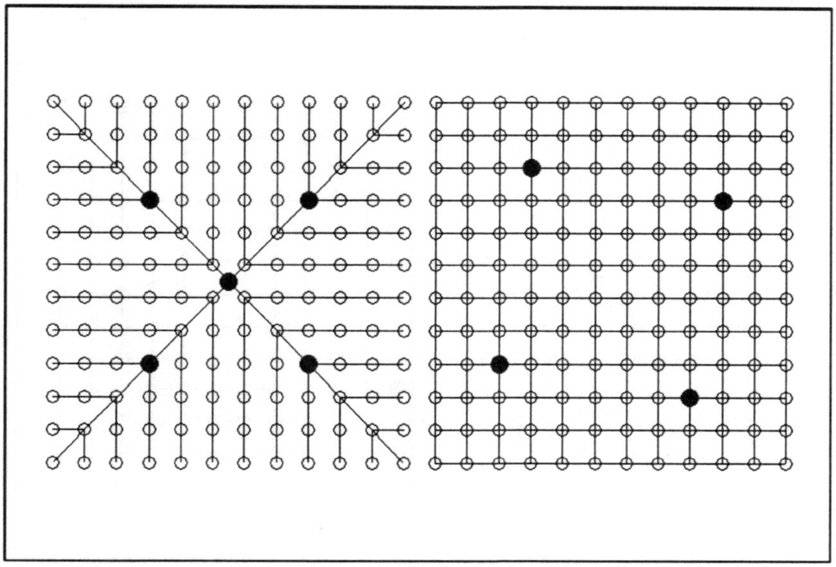

$F = 5000$

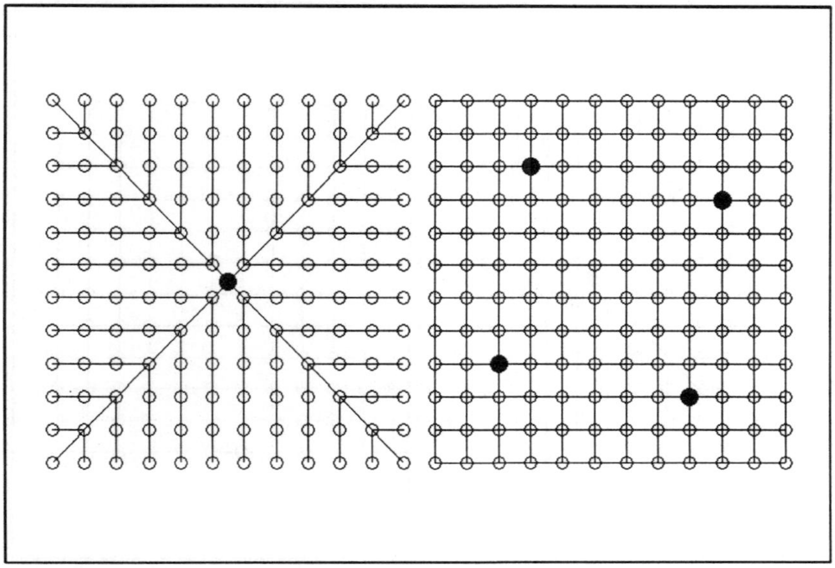

$F = 7500$

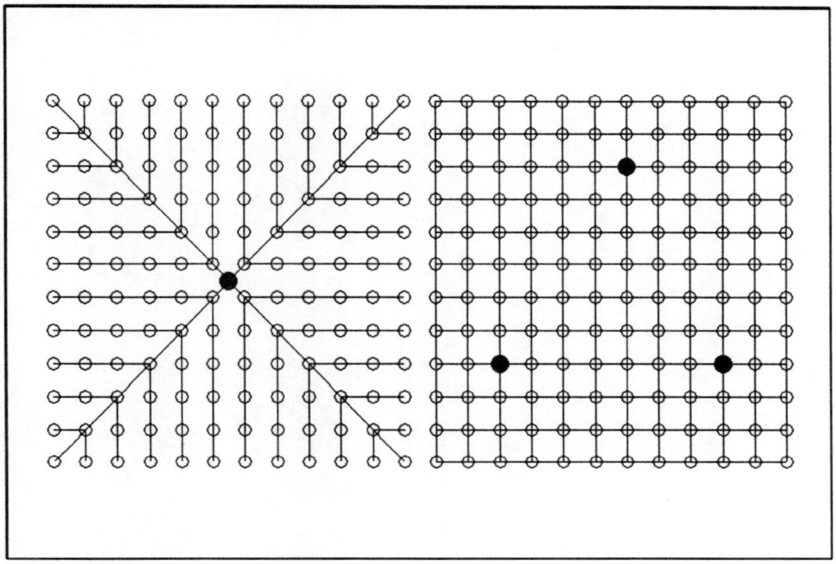

$F = 10\ 000$

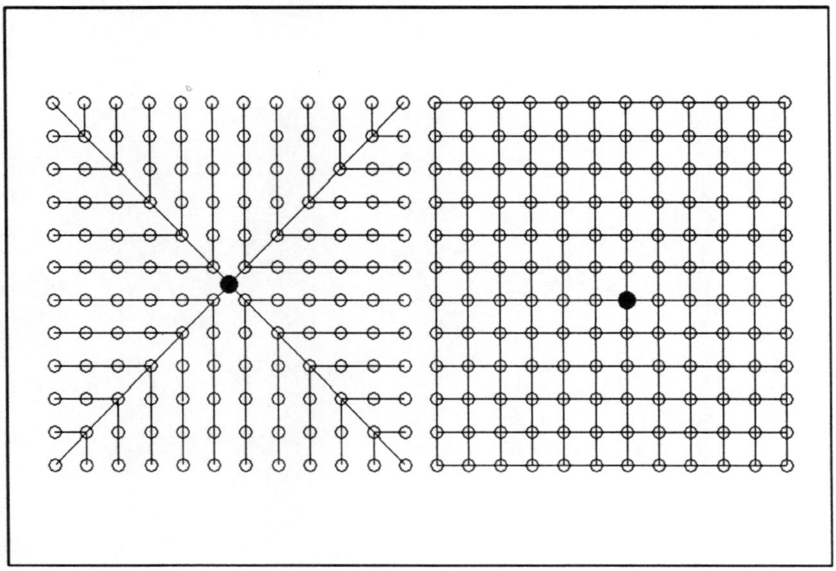

$F = 30\ 000$

Figure 4.8 Optimal locations for two autarkies and for different values of F

5. Optimal locations of human activities and the permeability of the border in a common market

This chapter is a continuation of Chapter 4 in that, given a common market with two economies characterised by their own transportation structure, we now look at the way the two economies are bound together: is the border characterised by many gates? Where are those gates located? Do these planning choices (number of gates, location of the gates, tariffs associated with the use of the gates) influence the location-allocation results? What happens if an extra high-speed connection is added between two regional centres?

This chapter is organised as follows: Section 5.1 gives an overview of the problem and makes a link with the existing literature. In Section 5.2, we consider the impact of the changes in the permeability of the border on the location-allocation results in a common market. In Section 5.3, we study the impact of a high-speed connection between the two regional centres on the spatial distribution of activities and on the structure and hierarchy of the production centres. In the last section, we summarise our main results and draw some policy implications.

5.1 INTRODUCTION

5.1.1 General Context

The mutual relationship between transportation and regional development is a matter of continuous concern in town and country planning (see for example Rietveld, 1989a; Nijkamp, Perrels and Schippers, 1995; Rephann and Isserman, 1994). In Europe, the introduction of important changes in inter-regional communication networks such as the French TGV connections (see Bonnafous, 1989; Klein, 1998; Troin, 1995), the Channel Tunnel (Vickerman, 1994, 1995 or Dundon-Smith and Gibb, 1994) or the planning of bridges in southern Scandinavia (Matthiessen, 1993) have given further stimulus to research in this field. Moreover, in a more distant future, the Maglev transport system will probably be capable of transporting passengers and goods at speeds that approach and perhaps even exceed those of modern jet transport in Europe; this could, by analogy, connect different cities and urban corridors into a 'Continental settlement super-cluster' in a so-called 'ecumenopolis' (Nakicenovic, 1995).

It is important to note that until now research on the influence of regional development on transportation flows has been more successful that on the reverse relationship. An increasing concern is now also voiced on the barriers in Europe caused by missing links or even missing networks (Nijkamp, Perrels and Schippers, 1995). Transportation policy favouring a free movement of persons and commodities in the EC is an imperative condition for a single market; the planning of trans-european networks witnesses this. The removal of barriers is of great importance for obtaining the highest possible dynamic integration effects from a network economy.

In the preceding chapter, we considered the case of a common market made out of two squared lattices linked together by a maximum number of gates: the number of gates was equal to the common number of horizontal links in both networks (fully integrated common market). Barrier networks are distinct from the above in that their structure is designed to impede or resist flow to certain areas for locations: obstacles to access are inherent in such networks. This is for instance the case of (1) national borders consisting of a physical barrier (river, mountain, and so on) with a limited number of bridges, saddles or tunnels; (2) limited customs with long waiting times for transportation or important trade-tariffs (for example between the former East and West Europe); (3) socio-political and economical isolation of many regions or nations (see the example of the Balkans in Petrakos, 1997). Moreover, barriers can be endogenous to lagging regions or exogenous if promoted by a centralised political structure, international constraints or by the nature of a core region's network (Suarez-Villa, Gaoutzi and Stratigea, 1992). It is generally recognised that national borders function as barriers in international interaction patterns. In the EC, good connections in the international infrastructure networks are considered to be a critical success factor in the distribution of economic activity (for example Bruinsma and Rietveld, 1993; Rietveld and Wintershoven, 1998). Hence, in Section 5.2, the preceding simulations (Chapter 4) will be pursued by analysing the effects of changing the number and the location of the 'gates' between two regional economies.

There are also ongoing debates within the European Union about the relevance of constructing a high-speed railway connecting the main conurbations of the Union (HST). The main motivation for such a policy is that, by improving global accessibility, it should both allow for the development of inter-regional trade and foster a better economic and social integration of European regions (van den Berg and Pol, 1998). In this perspective, it is important to determine access barriers and network structures that can introduce significant obstacles to passenger and commodity flows. This is also the opinion taken by the European Commission in designing its regional policy on the grounds that lagging regions have a level of infrastructure much lower than that observed in the leading regions.

In the past, urban clusters have condensed along important communication infrastructures such as railway lines with telegraph wires running next to them. Examples include the Tokyo – Osaka corridor served by the Shinkansen train, the Boston – New-York – Washington corridor linked by rapid rail and air shuttles and the Ruhr area traversed by intercity trains and highways. Yet, it is widely recognised that the provision of transport infrastructure does not necessarily attract activities in the regions affected by the new infrastructure (Offner, 1997), and therefore is not sufficient to promote regional economic development (for example Plassard, 1976; Rephann and Isserman, 1994; Vickerman, 1994a and 1994b), although some authors are less sceptical (Quinet, 1992; Rietveld, 1989a and 1989b; Rienstra et al., 1998). In particular, the decompartimentalisation of backward regions does not seem to trigger a process of local growth. Worse, such a policy might even be originate some economic decline. The waning of the Mezzogiorno that has followed the Italian unification is a good example.

The role of public capital, as emphasised in the endogenous growth literature, provides another possible justification for the development of transport infrastructure. According to Aschauer (1989), the provision of public capital (which here encompasses training, hospitals, sanitation, highways and transport) would have a significant, not to say major, impact on economic growth. And indeed, it seems very likely that investments in public infrastructure both create new resources and enhance the productivity of existing resources, even at the state level (Munnell, 1992). For example, in the US where streets and highways account for one–third of the total stock of public infrastructure, improvement in the transport system should foster more economic growth in the corresponding regions.

However, this is a policy recommendation in search of a theory since we do not know yet what is a good infrastructure policy in the context of a spatial economy (see Martin and Rogers, 1995, for a first attempt to evaluate the impact of infrastructure on the regional distribution of activities in a model of economic geography). Furthermore, subsequent empirical work has cast serious doubt on the conclusions obtained by Aschauer and others. In particular, the direction of causation is not clear since it is reasonable to expect a rich region to invest more in public capital (Duffy-Deno and Ebert,

1991). At this stage, it seems fair to say that the dominant opinion in the economics profession is rather mixed, especially regarding transport infrastructure whose impact is likely to be weaker than that of education or sanitation in public spending (Gramlich, 1994).

It should also be noted that investment in transport infrastructure has proven to be fairly ineffective in narrowing the gap between European regions while other policy instruments appear to be more efficient (de la Fuente and Vives, 1995; Martin, 1998), thus strengthening doubts about the global productivity of transport infrastructure. Note that assessing the impact of a transport infrastructure on regional development remains a hard task because this infrastructure is part of a much larger network which is itself in interaction with a hierarchical system of cities and nodes (Vickerman, 1994, 1995).

All of this does not mean, however, that transportation policy has no impact on the economy. Indeed, this policy has multiple facets and there exists a wide variety of approaches towards modelling infrastructure (Rietveld, 1989a and 1989b). Hence its impact could well be found somewhere else. It is our contention that the provision of transport infrastructure has more impact on the spatial distribution of activities than on macroeconomic growth. If this is so, it is then erroneous to expect regional or national growth to be directly and strongly affected by new transport infrastructure. Instead, as suggested by location theory (Beckmann and Thisse, 1986; Hurter and Martinich, 1988), one should first investigate what could be the impact of the morphology of a network on the spatial distribution of firms and households. To be sure, the geographical distribution of economic agents may in turn affect global growth but the nature of the relationship is still unclear. Although it seems reasonable to think of the spatial agglomeration of activities as the geographical counterpart of economic growth (Ottaviano and Puga, 1998).

In our opinion, any attempt made to assess the impact of a new infrastructure should account explicitly for the morphology of the network as well as for the nature (for example, the French TGV or the future Maglev system) and the relative position of the new piece of infrastructure (the cities which are connected by the TGV according to a particular scheme). Indeed, building new connections not only reduces transportation costs (broadly defined to include all impediments to interaction) between some places; it also affects (1) the accessibility of these places relative to others as well as among themselves, and (2) their position in the hierarchy generated by the network. This is precisely this kind of network effect we want to study here in the special case of a rapid train connecting two regional centres. So we do not deal with the possible impact of an entire TGV (HST) network (Vickerman, 1995). Specifically, in Section 5.3 we propose to study how the number and the locations of production facilities are affected by a major transformation in the transport system such as the construction of a high-speed connection.

5.1.2 Methodology and Expected Results

Since it is more than problematic to derive analytical results once locational decisions are treated as 0–1 variables, we have chosen to resort to computational methods which have been successfully developed in integer programming, to determine the configuration of production facilities. In particular, we focus – in this chapter as in the previous one – on the simple plant location problem whose purpose is to find the number and locations of facilities which minimise the sum of the fixed production costs associated with the setting up of facilities and of the transportation costs of users to the facilities to be located. This model is one of the prominent location problems studied in facility location analysis and very efficient algorithms have been designed to solve it (see Cornuejols, Nemhauser and Wolsey, 1990, as well as Labbé, Peeters and Thisse, 1995, or Daskin, 1995 for recent surveys). It is briefly described in Section 4.2.1.

As already discussed in the preceding chapter, the simple plant location problem focuses on the trade-off between fixed production costs and transportation costs. This trade-off is at the heart of many location models (Beckmann and Thisse, 1986; Mulligan, 1984) and may be encountered in planning a system of public facilities, such as schools or recreational facilities, as well as in the design of a production–distribution strategy by a private firm (Erlenkotter, 1977). Observe also that this trade-off is central in economic geography where it appears in the pioneering analyses developed by Christaller (1933) and Lösch (1940). On the one hand, the existence of scale economies at the firms' level is a critical factor in explaining the emergence of economic agglomerations. On the other hand, the need to interact among individuals and the corresponding congestion costs (defined broadly in order to include competition for land use), imply that all activities are generally not concentrated in one place. Consequently, the trade-off between scale economies in production and transportation costs is critical for the geography of human activities and so regardless of the particular institutional setting in which those activities develop (see Fujita and Thisse, 1996, for a more detailed discussion). This explains why we have chosen to work with a model based on that trade-off.

There are at least two more reasons for retaining the simple plant location problem. Firstly, for any given spatial distribution of demand, the solution to this problem gives the socially optimal configuration of facilities that minimises resource consumption. Hence it is likely that our results shed light on what could be the ideal network design policy. Secondly, to the extent that the market economy yields outcomes that are (more or less) socially optimal, the solution to the simple plant location problem may be expected to be a reasonable approximation of the market solution. In this perspective, the theory of contestable markets suggests that potential competition may well discipline market economies with non–convex technologies (Baumol, 1980). For our purpose, the work by Demange and Henriet (1991) is worth noting since they show the existence of configurations of firms in a spatial economy such that no consumer can be made better off by changing unilaterally the strategy of a firm. In other words, our optimal configuration could well correspond to a certain type of market equilibrium.

The simple plant location problem is – as in the previous chapters – applied to toy-networks, which provide approximations of real-world networks. In this chapter, we concentrate upon grid (rectangular) networks only because they do not introduce a priori any form of centrality or hierarchy inside the system of places connected by the network. We then solve many problems for various pairs of such networks, with several locations and numbers of gates between the two economies, as well as with and without a high-speed connection between the two regional centres, in the hope of uncovering the possible impact of such a connections on the global distribution of activities in the integrate economy. So, our approach differs here from that taken in the preceding chapter where we compared the impact of different network shapes on the location of facilities.

Unexpectedly, we will see here that the permeability of the border between the two regions has little impact on the location of human activities. Adding a TGV link between the two regional centres, we will also see that the relative value of fixed production costs to transportation costs is crucial in the determination of facility locations. Even less expected, perhaps, is the fairly strong instability observed in the structure of the locational pattern of facilities when the relative value of the fixed costs changes. This points to the same direction as recent models in economic geography focusing on the same trade-off (Krugman, 1991; Ottaviano and Thisse, 1998), although the results are not directly comparable because regions in those models have no spatial extension. Furthermore, a high-speed connection turns out to have a significant impact on locations when the following two conditions are met: (1) the saving in transport spending is substantial, and (2) the transportation costs are low relative to production costs. While the gateway between the two regional economies is likely to be the main production centre in the absence of a high-speed connection, we will find that such a connection tends instead to foster the development of one regional centre at the expense of the other when those two conditions are met. In other words, activities that are very sensitive to the time saved should be markedly affected by the building of a high-speed connection in that their locations would be concentrated into a single regional centre at the expense of the other. Although it is premature to conclude, it is worth mentioning that, since the Channel Tunnel has been built, the City of London has already attracted a lot of French financial operators who worked in Paris before. Simultaneously, one observes a growing number of British shoppers in Paris. This would be because the building of a high-speed connection primarily affects activities involving a specialised workforce. Alternatively, the supply of such a piece of infrastructure could also lead to substantial economies of transport density (Braeutigam, Daughety and Turnqvist, 1984) and is expected to produce similar effects.

Hence, Section 5.2 will study the changes in density and location of the gates between the two regional economies, and Section 5.3 will consider the effect of an extra high-speed connection between the two regional centres.

5.2 HOW FAR DOES THE PERMEABILITY OF THE BORDER INFLUENCE LOCATION-ALLOCATION RESULTS IN A COMMON MARKET?

5.2.1 The Studied Toy-Networks

The aim of this section is to assess the impact of the existing transport barriers on the functioning of the single market (Section 5.1.1). In the EC, for instance, many of the benefits of the single market programme stem from increased competition due to liberalisation of the regulatory framework and more efficient allocation of resources as a result of the removal of non-tariff barriers to free intra-EC trade. Bruinsma and Rietveld (1993) showed that national borders exert a barrier effect on international communication especially as far as rail and road traffic are concerned. With rail, two elements play a role: network densities on borders and the frequencies of international trains. Bruinsma and Rietveld limited themselves to the investigation of the impact of barrier effects on the accessibility measures.

In this section, we pursue the simulations performed in Chapter 4 by analysing the effects of changing the number of gates between the two regional economies or changing their location. What is the effect of limiting the number of physical gates between two regions on the optimal locations and allocation of human activities within those regions? Moreover, given a number of gates (or given a density of gates), do their locations affect the results of the location-allocation model? Hence, the role of barriers in communication networks and their impact on territorial and border region development is explored in this section. The hypothesis we want to test is the following: the morphology of the border influences the relations between two economies and hence the locations of the human activities.

Once again, we consider a discrete transportation space. The chosen regional lattice of points is a 15 × 15 squared lattice with 225 points, where each point has the same spatial environment as every other at the exception of the border points. Each point of the lattice is simultaneously a demand point and a potential location for a facility. Each point j is characterised by its coordinates and is linked by an edge to its closest neighbours. Each link has a length of 100 for horizontal or vertical edges. This lattice represents one regional economy. The number of edges and their spatial organisation define the transportation network. In this chapter, only one type of network is considered: the grid network. We know from previous work that such a network allows for more flexibility in location choices than, say, a radial network (Chapters 3 and 4). The requirement of the composite good in any point j is fixed and invariant with j, that is, the distribution of demand is uniform with $\delta_j = 1$ for all j. Two identical squared networks are then merged in order to see how the optimal locations in a common market depend upon the connections between the underlying regional networks; the two networks are denoted L and R. Several densities and locations of links between the two regional economies are now considered.

Figure 5.1 sketches some of the toy-networks that will be used in this section. Firstly, one can think of a variable number of horizontal links between the two regional economies, each link having a length of 100 as those belonging to the regional networks. Hence, the potential number of links or gates (noted here γ) between the two economies varies from 0 (the two lattices are totally independent) to 15 (full linkage used in Chapter 4). In this section, we consider values of γ respectively equal to 1, 3, 5, 8 and 15.

Secondly, it is important to see where those gates are located. Four different spatial distribution rules for the location of the gates are here considered: in the regular distribution, the gates are spaced regularly from top to bottom on the border (Figure 5.1, A). In the random distribution, a uniform random process chooses the index of the pairs of nodes at each end of the gate; two random sets are used in the following sections and are noted *random1* and *random2*. In the central distribution, all gates are located centrally (like parallel lanes of a motorway) (Figure 5.1, B). Finally, the concentration of all gates at the upper limit of the border has also been considered (Figure 5.1, C).

At the exception of Section 5.2.4, the length the gates (noted λ) is given and fixed to its 'natural' value ($\lambda = 100$): no tariff is applied.

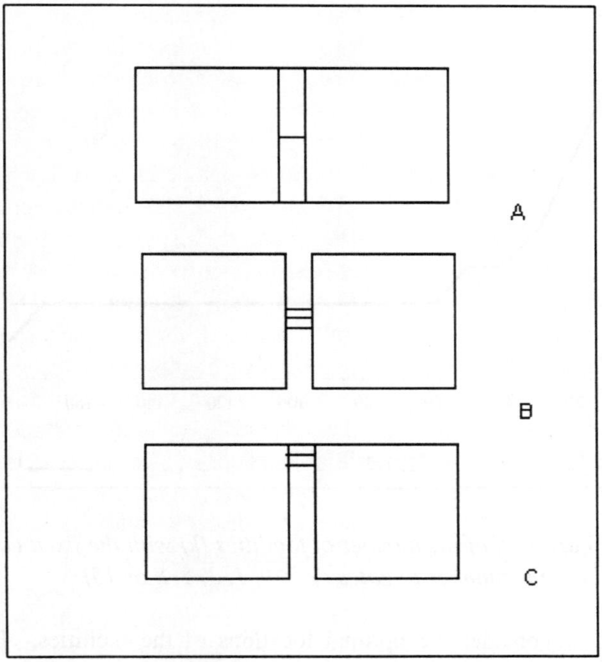

Note: A: 3 gates located regularly; B: 3 gates located centrally, C: 3 gates located at the upper limit

Figure 5.1 Examples of toy-networks used in Section 5.2

5.2.2 Sensitivity of the Location Results to the Number of Gates

Figure 5.2 gives the variation of the number k of facilities with the fixed cost F for various numbers of gates (γ) located regularly; this figure shows that whatever the value of γ, the number of facilities decreases when F increases. Variations with γ are small (a maximum difference of one facility) and only observed for values of F smaller than 60 000 EUR. Operationally, this means that in terms of number of facilities, there are very little changes in the location results with the number of gates between the two regions: multiplying or reducing the number of bridges or gates between two regions has little impact on the number of facilities located (everything else being equal: no congestion, no tariffs, no behavioural changes, and so on).

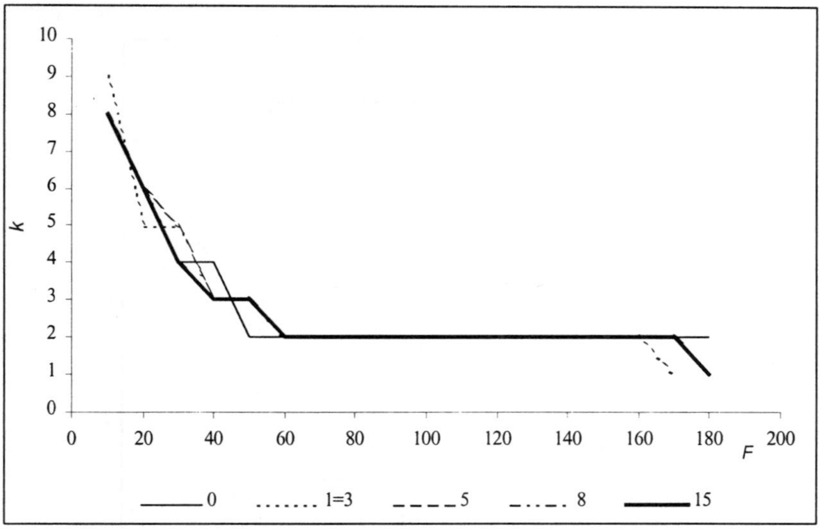

Figure 5.2 Variation of the number of facilities (k) with the fixed costs (F) and the number of gates (γ = 0, 1, 3, 5, 8 or 15)

Let us now consider the optimal locations of the facilities. Figure 5.3 illustrates the changes with γ in the locations of the concentrations of human activities when the gates are regularly spaced. γ, the number of gates, is here equal to 1 and 3 and k, the number of facilities, is equal to 9. Figure 5.4 is similar to Figure 5.3 but refers to values of γ varying from 5 to 15; in this case k is equal to 8. F is equal to 10 000 EUR in both figures. Note that the same conclusions are obtained with other values of F (not illustrated).

From these two figures, we can conclude that:

1. When γ is small ($\gamma = 1$ or 3), nine facilities are necessary to cover the common market; when γ increases ($\gamma > 3$) the gateway city does not play any structural role and less facilities ($k = 8$) are sufficient to cover the two regional economies.
2. The location changes observed with γ do not follow a well–defined spatial rule: small 'erratic' changes occur.
3. The locations of the activities further away from the border are less affected by the variations in the permeability of the border (γ) than those located closer to the border.
4. Following Taaffe, Gauthier and O'Kelly (1996, pp. 13–14), gateways describe nodes or centres that often serve as main entry and exit points for a region and at which the transhipment or freight-forwarding function is

prominent; these authors state that 'the role of gateways is more evident from relatively subjective historical investigation that it is from precise (accessibility) measurement' (p. 17). This is confirmed by our results: gateway cities do not induce a strong and unambiguous attraction force for the location of facilities except for small values of γ ($\gamma = 1$ or 3).

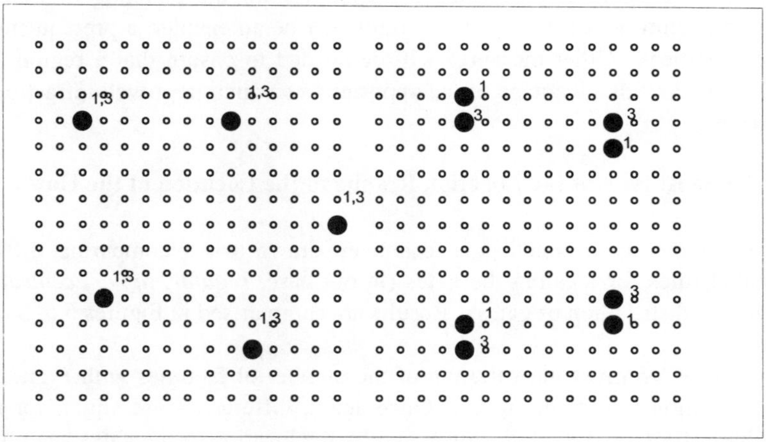

Figure 5.3 *Optimal locations when γ (number of gates) takes respectively the value of 1 and 3 (F = 10 000; gates regularly spaced)*

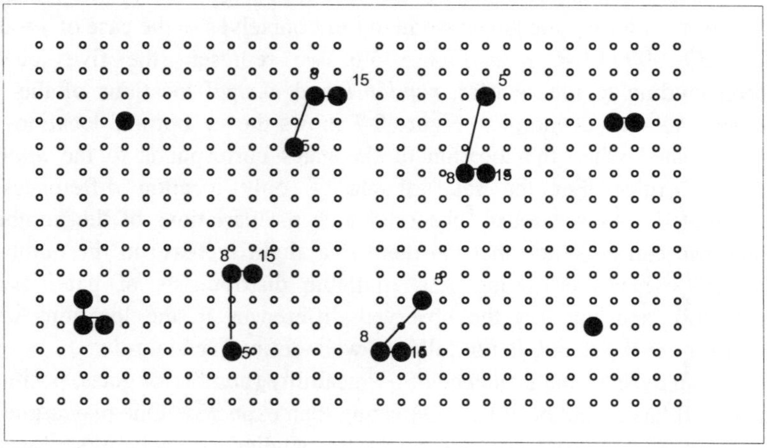

Figure 5.4 *Optimal locations when γ (number of gates) takes respectively the value of 5, 8 and 15 (F = 10 000; gates regularly spaced)*

On average, small changes are observed with γ – changes are much slighter than expected: whatever the density of gates, the number of facilities to be located is almost the same and the locations generally look alike. This means that the permeability of the border has little impact on the location results within the two regional economies, confirming Vickerman's (1995, p. 251) conclusion: 'One of the principal difficulties for the planning of transport infrastructure is that transport can only ever be an enabler, a prerequisite for development. Other measures will be needed to ensure that a region … is able to take full advantage of the opportunities which improved infrastructure creates'.

5.2.3 Sensitivity of the Location Results to the Location of the Gates

Almost the same conclusions can been drawn when comparing different spatial rules for locating the gates (in our case, *regular*, *upper*, *central* and *random* distribution of gates). Results are summarised in Figures 5.5, 5.6 and 5.7.

Figure 5.5 gives the variation of the number of facilities with F and with the location rule of the gates. Once again, differences are slight, for most values of F ($F < 160\,000$). Surprisingly, the location of the gates has little or no impact on the number of optimal facilities. This result is, however, to be interpreted within the limits of the conducted experiments: we did not investigate all combinations of locations of gates, nor all spacing of demand nodes.

Figures 5.6 and 5.7 compare the locations suggested for five different ways of positioning the gates; we here limit ourselves to the case of $\gamma = 3$ and $F = 20\,000$ EUR. In Figure 5.6, we represent the five locations corresponding to the *regular*, *random1* and *central* locations of the gates between the two economies; Figure 5.7 shows the six optimal locations (k = 6) suggested when the location of the gates corresponds to the *upper* or *random2* rules. For the sake of clarity, only location differences are represented. As previously observed with the variations of the number of gates, we can conclude that (1) there is a slight increase in the number of facilities when considering less equitable distributions of gates (*upper*, *random2*), and that (2) the observed differences in the locations of the facilities are slight and do not follow a well–defined spatial rule.

The analysis of the influence of permeability (number of gates, position of the gates) has in fact been less interesting than expected. One may argue here that this comes from the choice of the studied toy-networks, but other simulations were performed on larger (and smaller) sets of points leading to the same type of results and conclusions.

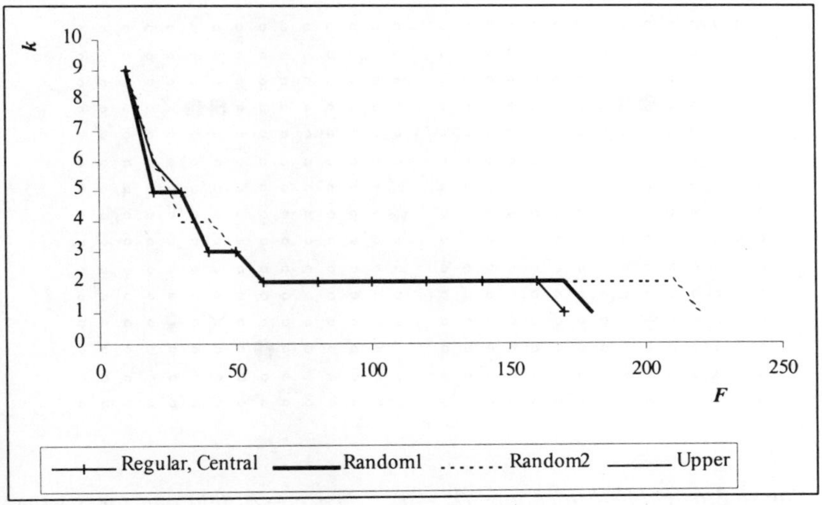

Figure 5.5 Variation of the number of facilities (k) with the fixed costs (F) and the type of location of the gates (γ =3)

In the following section, the length (λ) of the gates varies in order to show how far it influences the results of the optimal location-allocation models.

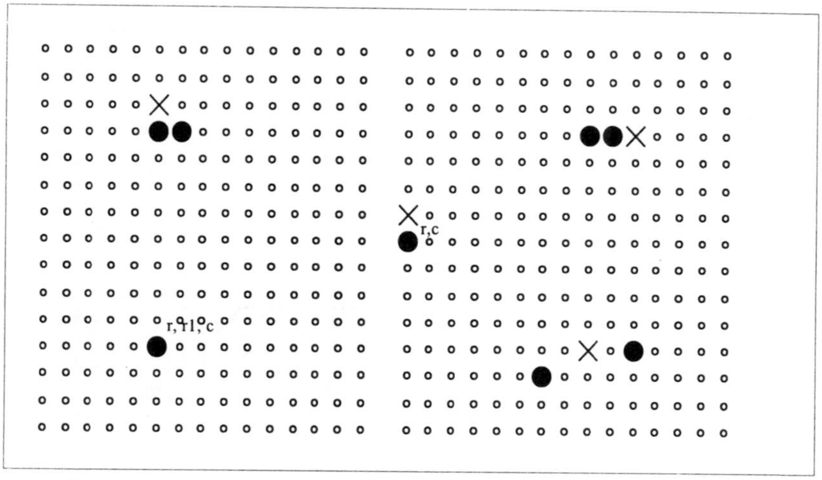

Figure 5.6 Optimal locations (k = 5) when F =20 000 and for 3 ways of locating the gates (black dot = central gates (c), grey dot = regular gates (r); × = random1 gates (r1))

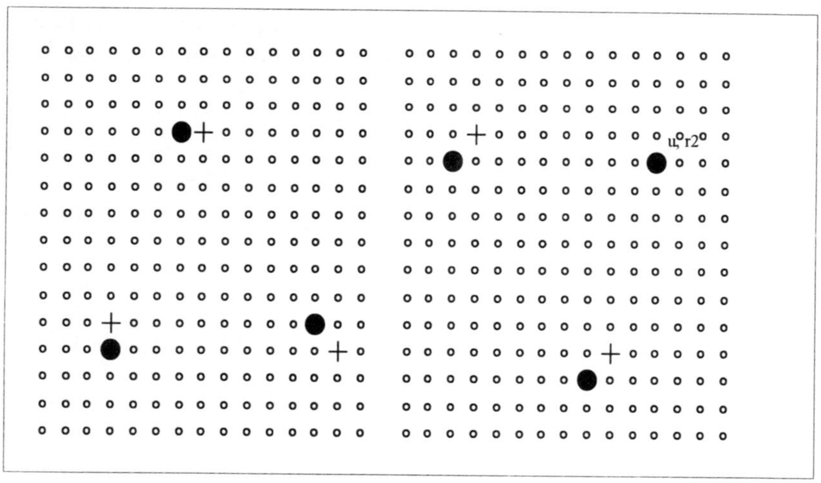

Figure 5.7 Optimal locations (k = 5) when F =20 000 and for 2 ways of locating the gates (black dot = upper gates (u), + = random2 gates (r2))

5.2.4 Sensitivity of the Location Results to the Length of the Gates

Given the preceding results and the numerous illustrations drawn – but not reported here – we here limit ourselves to the case of one gate ($\gamma = 1$) and one type of location (*central*), but we now let the length of the gate (λ) vary from 0 (no gate, two autarkies) to its 'natural' Euclidian length of 100 and up to a value ten times as high ($\lambda = 1000$). In doing so we want to understand the potentiality of the gateway city and the effects of taxes (for example tariffs, congestion that are expressed by λ) applied at the gate on regional development differentials.

As expected, the number of facilities decreases when F increases (not illustrated). As previously mentioned, slight differences are observed for medium and small values of the length of the gate (a maximum difference of one facility). One difference with λ is however interesting to mention: activities are concentrated in a unique production centre ($k = 1$) for values of F greater of equal to 150 000 EUR when λ is small. When we greathy increase the length of the gate ($\lambda = 1000$) these fixed costs are three times as large ($F = 400\ 000$) before the activities are all concentrated in a unique agglomeration centre ($k = 1$).

Figure 5.8 gives the optimal locations for a medium value of F ($F = 30\ 000$) and shows that :

1. No link between two economies ($\lambda = 0$) or a very large link ($\lambda = 1000$) gives the same location-allocation results: a spatially balanced configuration of activities, with no hierarchy of the centres. Note that when there is no link between the two economies ($\lambda = 0$), there are many alternative solutions; only one is represented here.
2. When a link exists between the two economies and when this link is short ($\lambda = 100$), the gateway city develops and extends its market area in both regional economies at the expenses of the other locations. This is the only case where a gateway location appears.
3. Weighting ($\lambda \geq 100$) of the length of the gate between the two economies generates a hierarchy between the central places. If we measure the hierarchy by dividing the predicted size (allocation) of the greatest site by the total allocation of the smallest site for all λ values, we see that this ratio varies from 1.00 ($\lambda = 0$) to 1.37 ($\lambda = 100$), 1.92 ($\lambda = 250$), 1.10 ($\lambda = 500$), 1,18 ($\lambda = 750$) and back to 1.00 when $\lambda = 1000$.

$\lambda = 0$

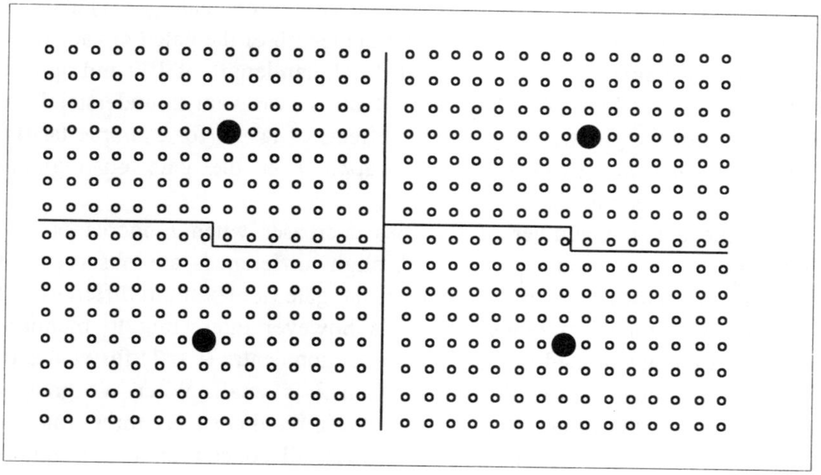

$\lambda = 100$

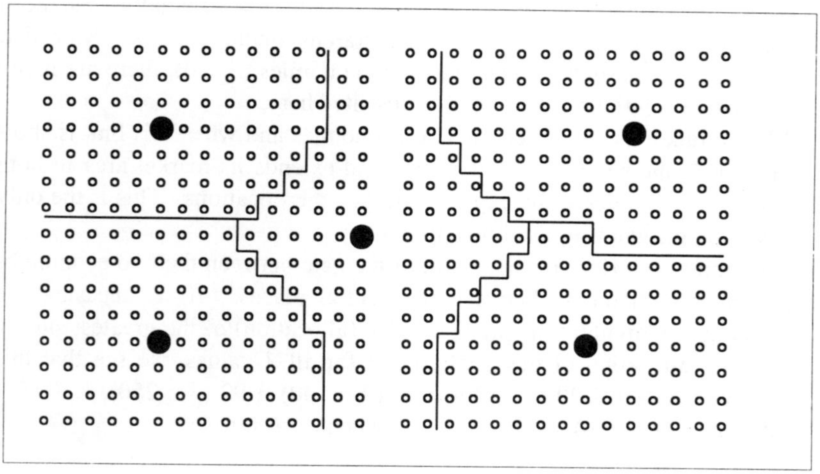

$\lambda = 250$

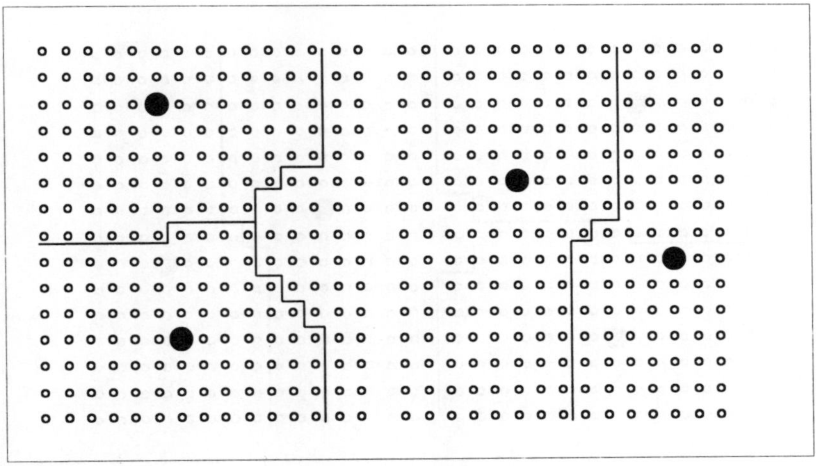

$\lambda = 500$

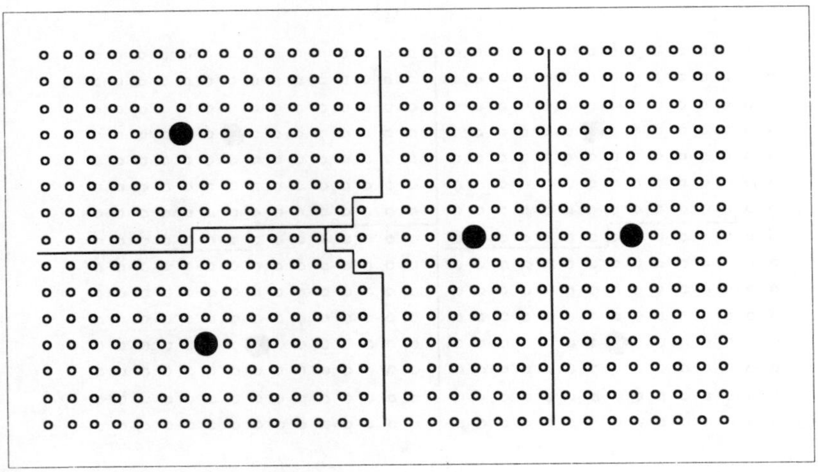

$\lambda = 750$

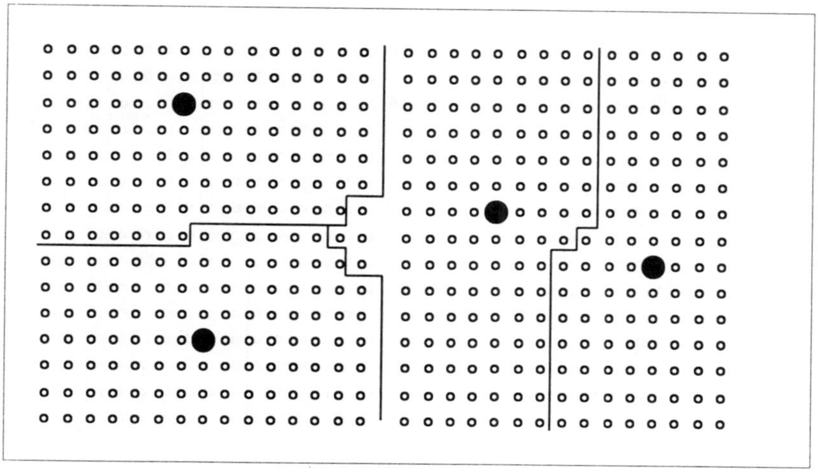

$\lambda = 1000$

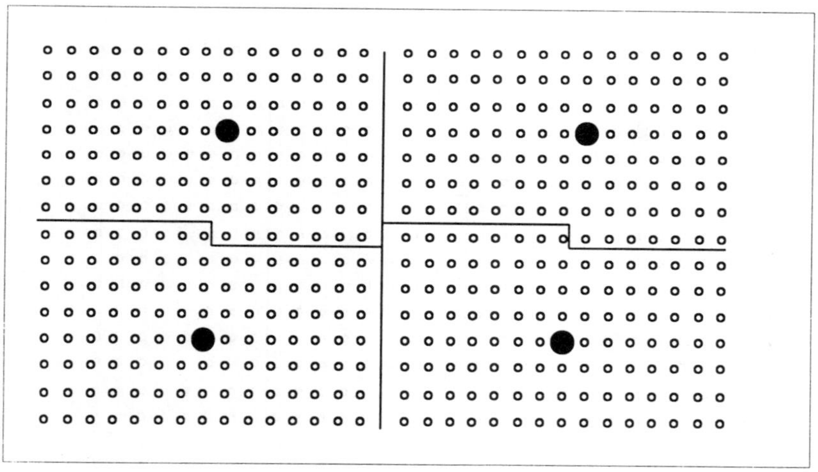

Figure 5.8 Variation of the optimal locations and allocations with the length of the gates (λ) for a given fixed cost (F = 20 000 EUR)

5.2.5 Some Preliminary Conclusions

At this stage of the analysis, some conclusions can be drawn about the effects of the permeability of the border on the location of activities within a common market. Firstly, we see that changes in location and allocation are slight compared to those expected or those discussed in the preceding chapters. Changing the shape of the transportation infrastructure in an economy (for example by adding a ring road) has much more impact on the SPLP results than adding some links at the border between the two economies. Hence, the shape of the border has little impact on the location results. Secondly, the length, the weight associated to a unique gate does have dramatic consequences on the location results especially when it is important (congestion, high tariffs). This is important in land–use planning and confirms existing cases (for example the former border between West and East Germany). In other words, the location of the gate has less importance than its 'length'. Thirdly, it is important to mention that location-allocation changes are hardly spatially predictable: they look erratic and do not follow a well-defined spatial rule; this is also of prime importance for decision–making; the sensitivity of the results is slight and unpredictable (see also Section 5.3). Finally, thinking to the EC example, we know that the removal of barriers is of great importance for obtaining the highest possible dynamic integration effects from a network economy. The partial or total suppression of the physical borders does not seem to induce dramatic consequences on the location of human activities. However, culture and language barriers still remain and the harmonisation of fiscal systems or the labour market legislation are far from being effective and can have much more impact on location decisions (see for example Nijkamp, Rietveld and Salomon, 1990). We here limited our simulations to the physical barriers only. The non-physical aspects of network planning should receive more attention in future studies and policies.

5.3 HOW FAR DOES THE EXISTENCE OF AN INTER-REGIONAL HIGH-SPEED CONNECTION AFFECT LOCATION-ALLOCATION RESULTS IN A COMMON MARKET?

In this section, we consider a single central standard link ($\gamma = 1$) between the two regional economies but also add an extra high-speed connection between the two regional centres in order to see how far this infrastructural change affects the locations of the regional activities.

5.3.1 The Experiments

As in Section 5.2, two 15 × 15 squared lattices of points are used; they are once again denoted L and R. Both are characterised by a grid network and the distribution of demand is uniform, with $\delta_j = 1$ for all j.

According to the results of the preceding section, we have chosen to keep matters simple by focusing on a single central link (such as a bridge, a tunnel or a saddle) between the two lattices ($\gamma = 1$). We then imagined that the centres of the two regional economies were connected by a high-speed train, which makes the two centres closer than by regular transportation means. Examples include the Eurostar between London and Paris or the Thalys between Brussels and Paris, as well as the Shinkansen between Tokyo and Osaka in Japan (see for example Murayama, 1994). This is the point we want to address in this section by simulating different locational configurations, corresponding to different values of the length associated with the link between the two regional centres. By reference to the current European policy, this link is here called the TGV link whose transportation cost is denoted by τ. Two situations are then compared: (1) the two economies L and R are connected by a standard link through a single central gate, which will be used as our benchmark, and (2) L and R are bound by both a central gate and a TGV link from centre to centre (see Figure 5.9 for a diagrammatic representation).

We also investigate the question of dense versus sparse networks by merging two networks, one of which is described by a 15 × 15 lattice and the other by a 5 × 5 lattice. In so doing, we want to control for the country's economic size effect. Hence, we assume that the global level of demand is the same in both economies in order to isolate the sole impact of the density of the network. This implies that the local demand in the sparse network is higher ($\delta_j = 9$) than in the dense network ($\delta_j = 1$)

In each experiment, the SPLP is run. For each selected value of the transportation cost along the TGV link, the model is applied several times, one for each selected value of the fixed costs. We assume that fixed costs are equal across locations in order to control for the role of differential factor endowments. In our experiments, this parameter, denoted F, varies from a floor value of 5000 EUR to a ceiling value of 170 000 EUR. The value of F thus expresses the (relative) importance of increasing returns in production. Furthermore, transportation costs by regular transportation means are linear in distance while marginal production costs are normalised to zero. This implicitly means that custom duties are set to zero between the two countries, as it should be within a common market. Transportation cost along the TGV link varies from τ=1500 EUR (that is, the amount paid for using the regular network) to τ = 300 EUR. These values have been selected because they generate a broad pattern of findings.

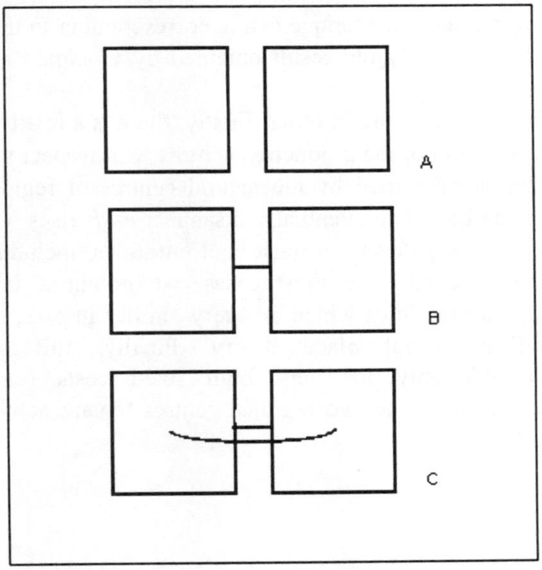

Figure 5.9 Schematic representation of the studied networks

5.3.2 A TGV Link between Two Economies with Equal Densities

No TGV link, one central standard link (the benchmark)

We first consider the case of two identical networks in terms of shape and density connected by a standard central link only (Section 5.2). We see from Figure 5.10 that for F = 5000 EUR the system of facilities is very close to the hexagonal pattern of central place theory. Space is evenly covered in that each facility has approximately the same market area in terms of shape and size. Both regional centres of L and R together with the gateway (or bottleneck) city are among the selected locations. When F = 10 000 EUR, the main striking feature is the fact that the two regional centres vanish from the map, whereas the peripheral centres of each region and the gateway remain active. When F rises to 20 000 EUR, the structure of locations is similar but with less facilities and there is a hierarchy among the facilities. The bottleneck city has now a significantly larger market area than the four secondary production centres which are (more or less) identical in shape and size. A further increase of F to 40 000 EUR brings back a regular covering of space by three places, including the gateway and two production centres located slightly away from the natural regional centres of L and R. For F =

60 000 EUR, there are only two locations corresponding exactly to the regional centres of *L* and *R*, while the gateway disappears as a production centre. Finally, for a very large increase in fixed cost (F = 170 000) production is concentrated in a unique centre corresponding to the bottleneck city. This confirms a fairly old result obtained by Goldman and Witzgall (1970).

The following comments are in order. Firstly, there is a relative instability in the spatial distribution of the production centres with respect to changes in fixed costs. This is magnified by the natural centres of regions *L* and *R* which vanish, come back and eventually disappear as *F* rises. On the other hand, we encounter the gateway in most configurations, including the most concentrated one. Secondly, in most cases, we obtain a fairly regular covering of space by facilities which are very similar in size, a result that seems to confirm central place theory. Finally, full geographical concentration occurs only for very high fixed costs (or very low transportation costs) since the two regional centres remain active from F = 60 000 to F = 160 000.

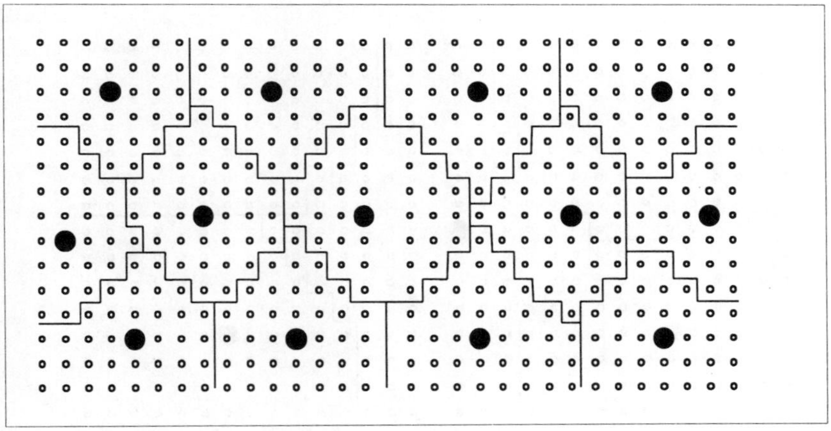

$F = 5000$

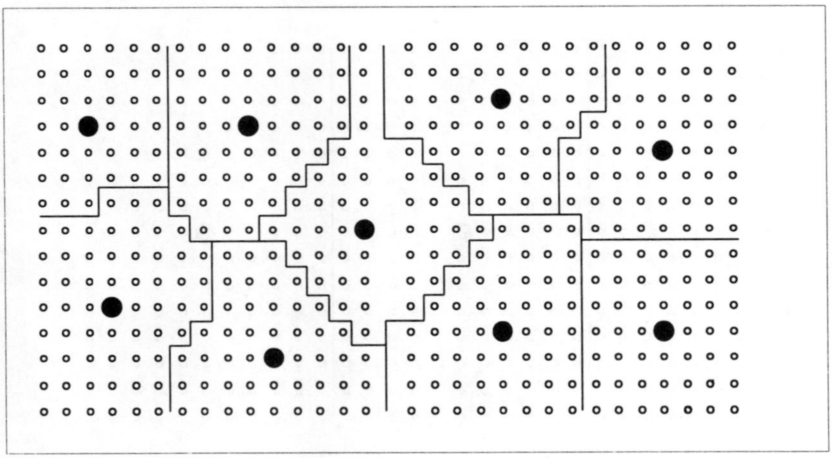

$F = 10\,000$

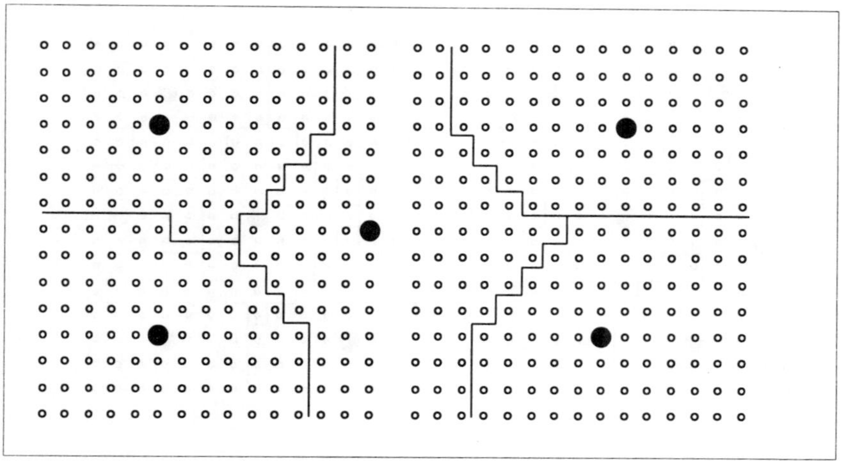

F = 20 000 or 30 000

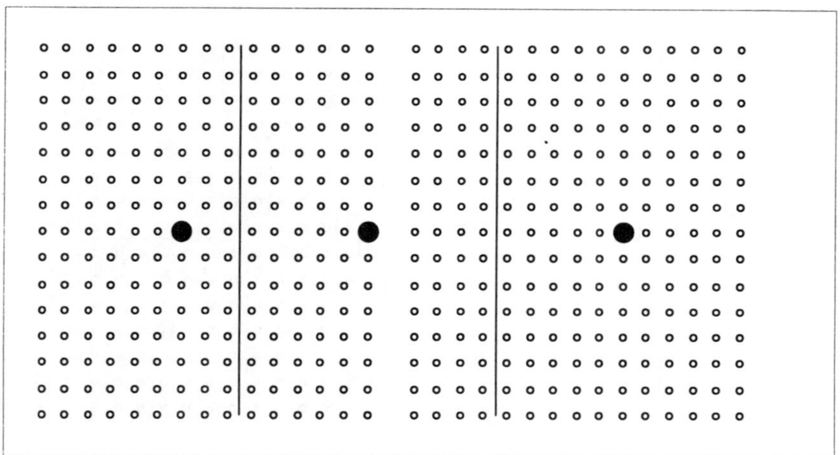

F = 40 000 or 50 000

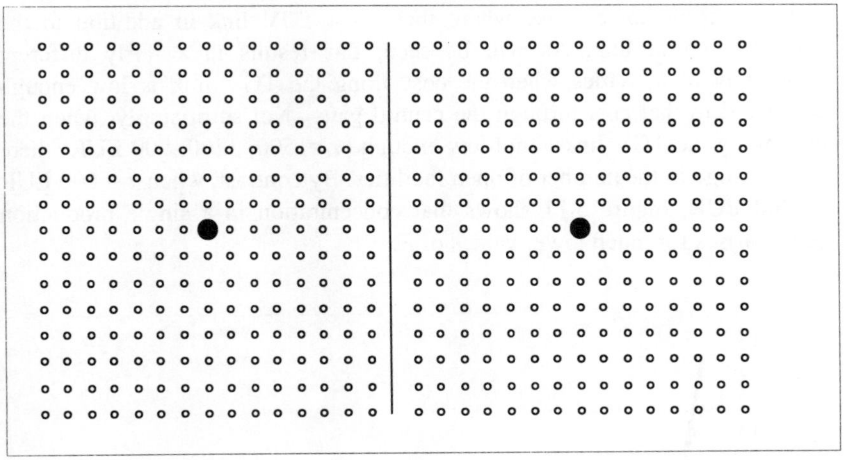

F = 60 000

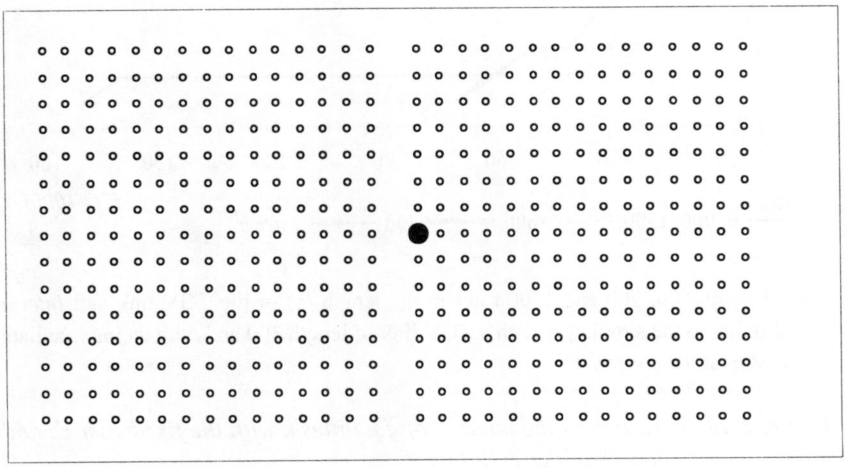

F = 170 000

Figure 5.10 *Optimal locations and allocations for different values of F when the two economies are bound together by a standard central link*

One TGV link, one central standard link

We now come to the case where there is a TGV link in addition to the standard central link. As will be seen, this results in a fairly different geography of activities when the cost along the TGV link is low enough compared to that cost through the central gate. Not surprisingly, when the cost along the TGV link is not low enough (τ =1500; 1200; 900 EUR) there is no change in the number of open facilities. By contrast, when τ = 600 EUR or 300 EUR, Figure 5.11 shows that concentration in a single production centre arises for much lower values of *F*.

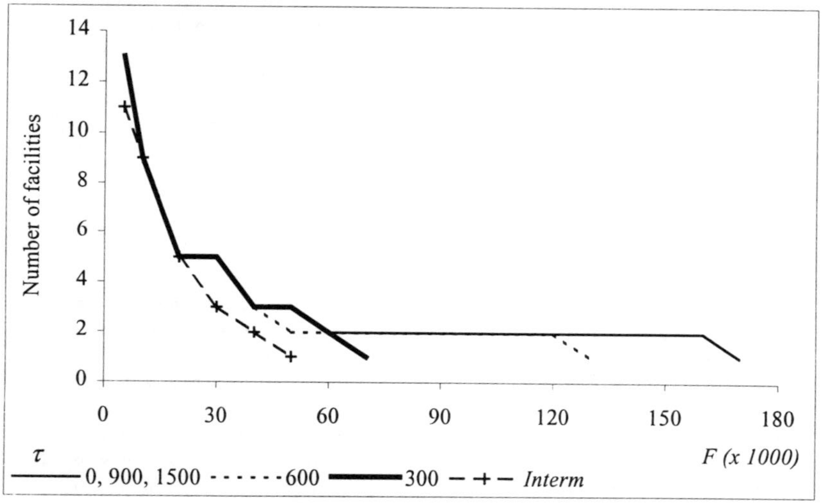

Note: 0, 300, 600, 900 and 1500 refer to the length (τ) of the TGV link and *Interm* refers to the simulation with a TGV link of length 300 and with an intermediate stop at the border.

Figure 5.11 Variation of the number of facilities k with the fixed cost F and the length of the TGV link τ

As shown in Figure 5.12, this is because a single facility set up at one regional centre allows for the provision of the other region at low transportation cost through this region's natural centre. Figure 5.12 reveals how the locations and market areas of the production centres vary with *F*. For very low fixed costs (*F* = 5000), the locational pattern of facilities is very similar to that obtained in Figure 5.10, thus implying that the TGV link has a negligible influence. On the other hand, things change when *F* = 10 000 EUR. The regional centre of, say, region *L* now supplies sites situated in the vicinity of the regional centre of region *R* which is no longer a production centre, albeit that the market area around the *L*-centre is smaller than that around the *R*-centre. In other words, one regional centre is swallowed up by its competitor, but its periphery still accommodates production centres. Another major change occurs for this value of the fixed costs in that the gateway no longer accommodates a facility.

Things change again dramatically when $F = 20\ 000$ EUR. Both regional centres disappear as production centres whereas the bottleneck city emerges again with a market area much larger than those of the peripheral centres. In a sense, the gateway together with the peripheral centres joins efforts to eliminate the regional centres. However, there is a hierarchy since the bottleneck plays the role of a primatial centre. When $F = 40\ 000$ EUR to 60 000 EUR, the configuration is similar to that obtained in Figure 5.10 without the TGV link. As mentioned above, in the presence of a TGV link, a regional centre becomes the only production centre as soon as $F = 70\ 000$ EUR, while the two regional centres would remain active without such a high–speed connection for larger and equal values of *F*. Hence, the TGV link allows one regional centre to capture all activities for intermediate values of F and eliminates the gateway as a production centre.

As observed in the absence of a TGV link, there is instability in the structure of locations according to the values of fixed costs. In particular, there is no monotonicity in the role played by the regional centres and by the gateway. This implies that it is hard to make predictions about the impact of the construction of a high-speed connection between regional centres on locations. Indeed, the emerging pattern of activities turns out to heavily depend on the relative importance of scale economies in production with respect to the decrease in transport spending made by using the TGV link. It is also worth noting that one regional centre takes over the other for very different values of *F* (10 000 and 70 000). This effect is generated by the presence of the TGV link and seems to confirm the observations discussed in the introduction.

So far, we have assumed that the TGV link connects directly the two regional centres without any intermediate stop; let us here cite the examples of the TGV-bis stations in France: Vendôme on the Paris–Tours line, Haute-Picardie on the Paris–Lille–Brussels line, Le Creusot or Mâcon-Loché on the southern line and some others planned on future lines (see for example Auphan, 1992, or Troin, 1995). As discussed in the introduction to this chapter, it is a matter of debate in Europe as to whether or not one should introduce intermediate points; Troin (1995) mentions that if many urbanistic and economic effects were expected, up to now none are observed. In our context, the gateway is an obvious candidate. In other words, we assume that the transportation costs between each regional centre and the gateway are equal. The number of operating facilities as a function of F is described in Figure 5.11. Furthermore, the change in the locational pattern is fairly dramatic: the bottleneck city is always a production centre (not illustrated). In particular, when increasing returns in the production sector are sufficiently strong (F =50 000), the gateway turns out to be the only production centre within the integrated economy. This provides a rationale for the claim made by several European cities (such as Turin in Italy or Lille in France) to become regional stops in the European high-speed railway system.

$F = 5000$

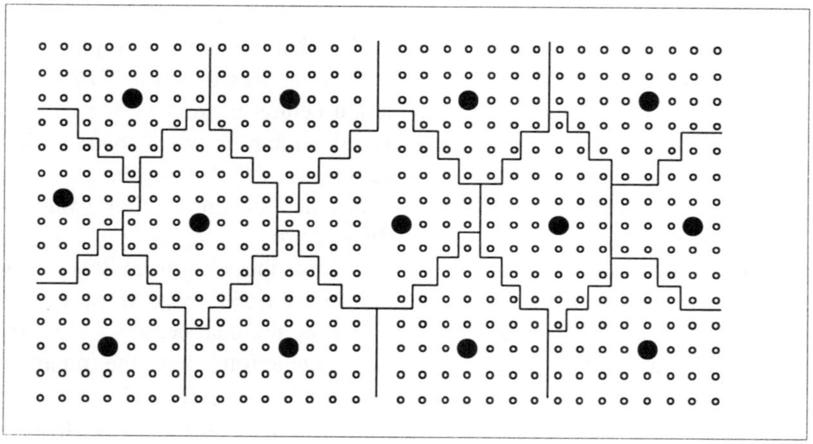

$F = 10\ 000$

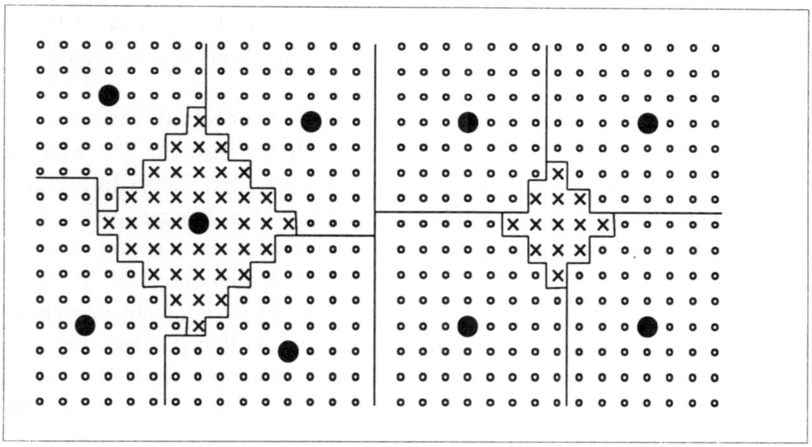

F = 20 000 and 30 000

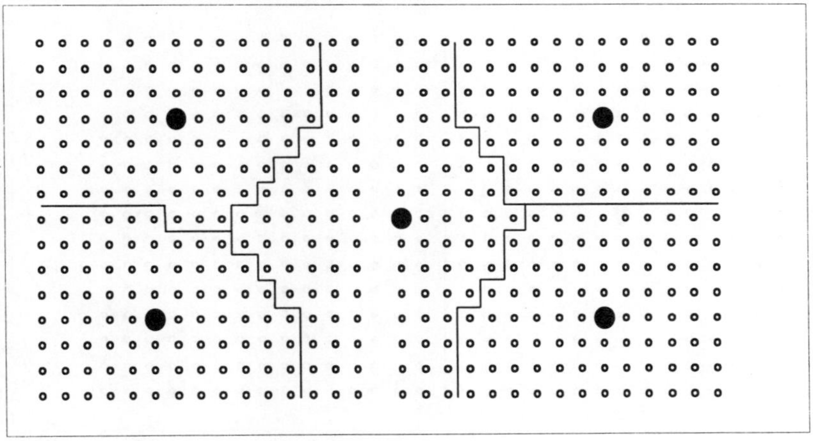

F = 40 000 and 50 000

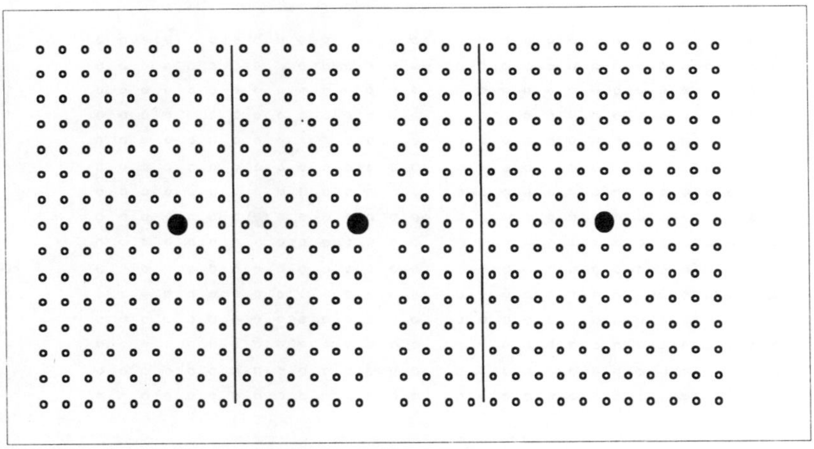

$F = 60\,000$

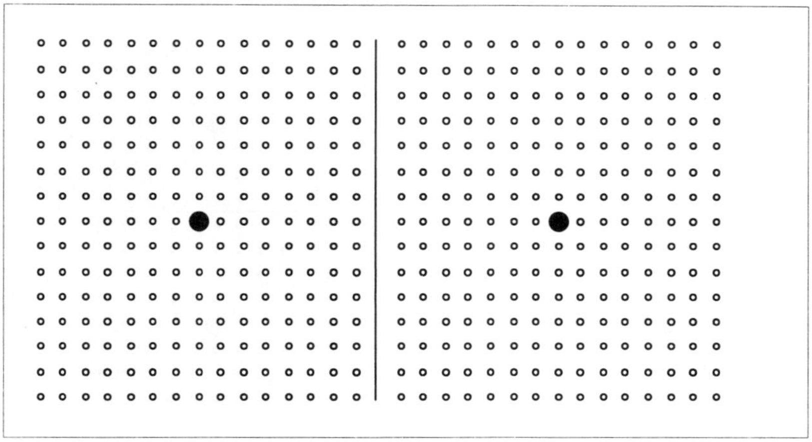

$F = 70\,000$

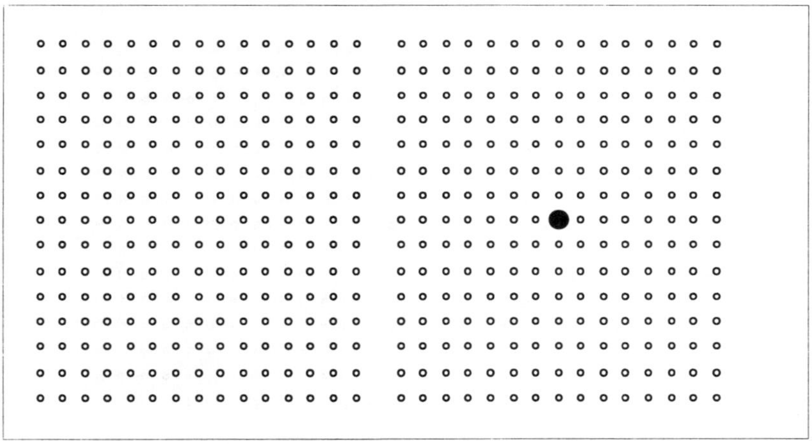

Figure 5.12 Optimal locations and allocations for various F when a TGV link ($\tau = 300$) is constructed between the two regional centres

5.3.3 TGV Link between Two Economies with Different Demand Densities

We now consider the impact of the TGV link (with $\tau = 300$) connecting a dense (L) and a sparse (R) network, but where regional demands are equal in both regions. The configurations are represented in Figure 5.13 for the values of F used in the foregoing figures. There are several differences with the case described previously. Firstly, as expected, there are more facilities in the dense network economy. Secondly, we observe the emergence of a single production centre in the sparse network economy from fairly low values of F (20 000 EUR). This location remains active for higher fixed costs and becomes eventually the only production centre within the two regional economies (60 000). Thirdly, the market areas of the facilities located in region R cover an increasing part of region L. For some values of F (10 000 and 20 000), the market area of the main centre in region R is formed of two pieces. Again, this is because the points close to the natural centre of L are now closer to the natural centre of R. When F increases, the penetration of region L by the centre of region R keeps rising until it covers the entire area under consideration. Lastly, the difference in network density acts in favour of the densely equipped region for low values of F because it accommodates more facilities, but this seemingly important comparative advantage is flipped in favour of the sparsely equipped region when increasing returns in production are strong enough. A corollary of this observation is the minor role played here by the gateway.

The fact that demand is less dispersed in region R leads to a faster concentration of activities within this region. This in turn allows the corresponding centre to exploit more the benefit associated with the existence of a TGV link to serve an increasing number of local markets in the other region. This situation is here inherently asymmetric because of the difference in network densities.

$F = 5000$

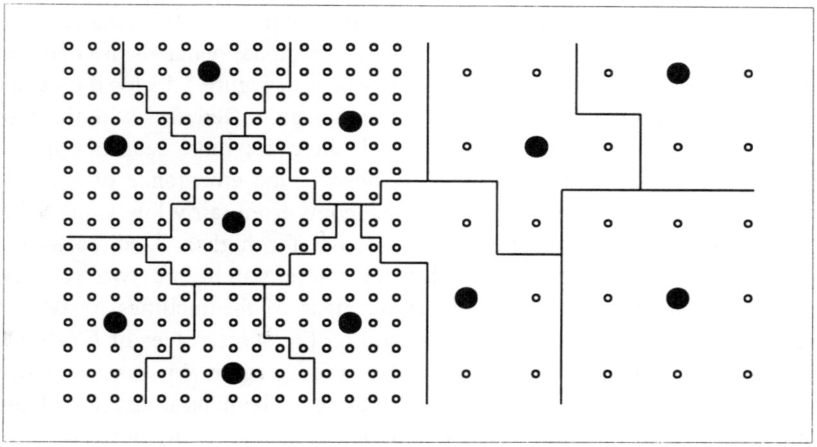

$F = 10\,000$

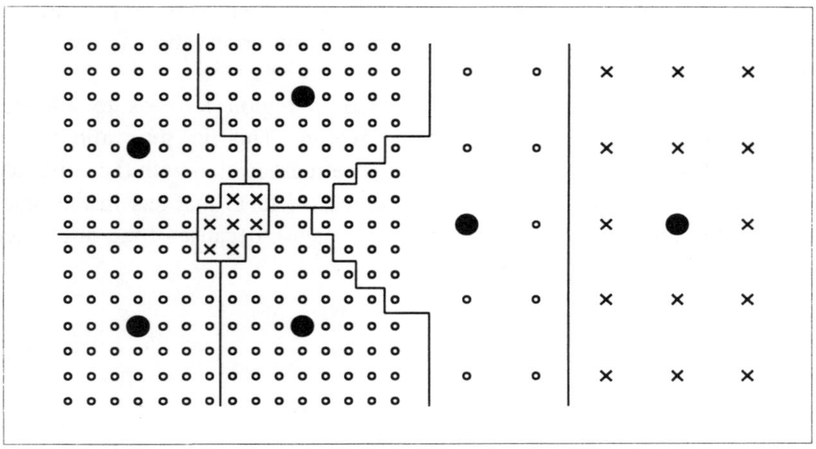

$F = 20\ 000$

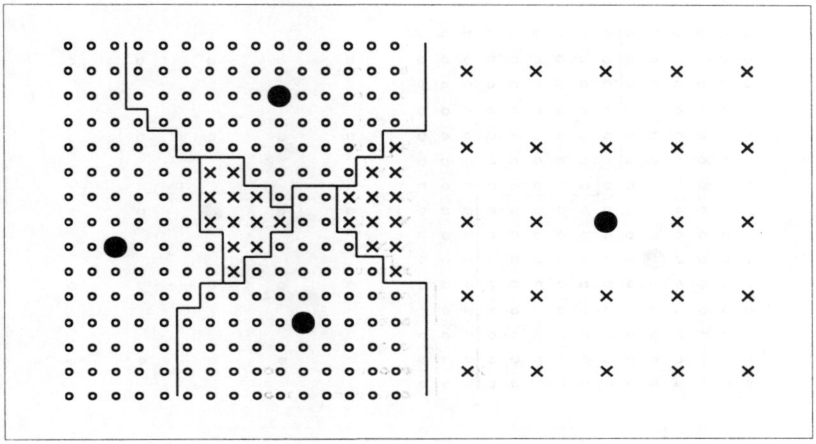

$F = 30\ 000$

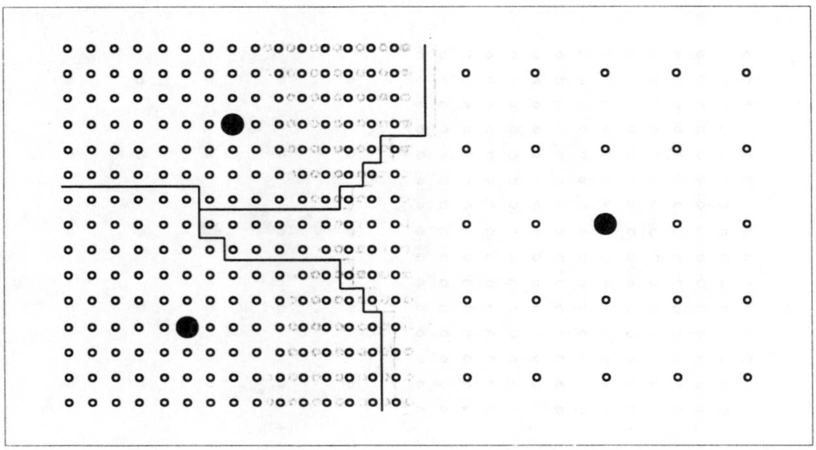

F = 40 000

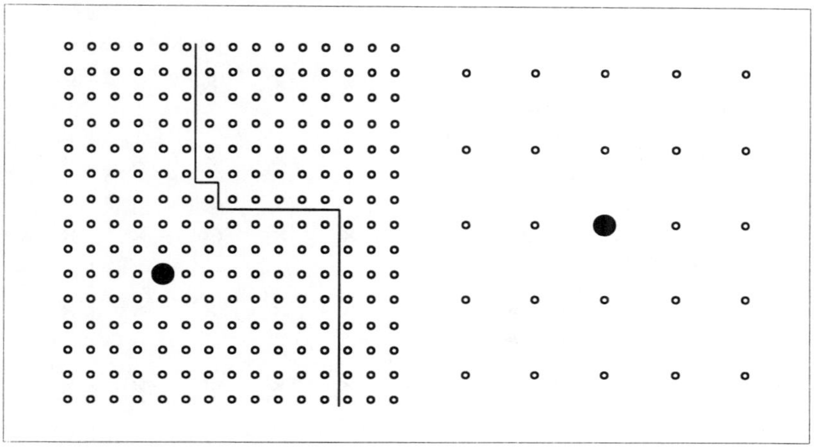

F = 60 000

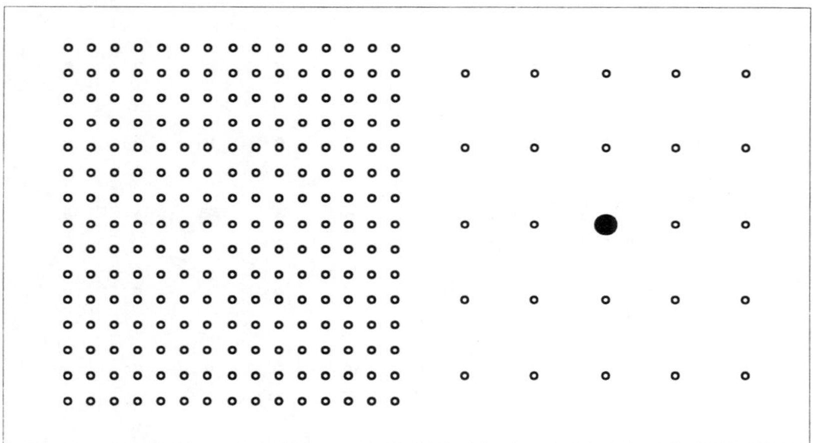

Figure 5.13 Optimal locations and allocations for selected F values when a TGV link with τ = 300 is constructed between the two regional centres and when one network is less dense than the other.

5.4 CONCLUDING REMARKS AND POLICY IMPLICATIONS

It is our belief that obtaining general results regarding the impact of transport networks on the location of facilities is highly problematic because of the combinatorial nature of the problem, even in the case of toy-networks. However, we believe that the kind of experimental approach followed in this chapter leads to interesting results because it relies more on what we know from spatial microeconomics. In this respect, using toy-networks, we have been able to study the impact of the construction of additional links on the location of activities. This is especially relevant when several economies are merged in a common market because national transport policies tend to neglect the global impact of international connections.

Our simulations have revealed some structural instability of the locational pattern as the value of the fixed production costs rises relative to that of the transport costs. This seems to confirm the scepticism of some transportation analysts in that it appears to be hard to predict what the final impact of a real-world infrastructure will be. Yet, the building of a high-speed connection between regional centres gives rise to some distinctive features (Quinet, 1998). Firstly, it does affect the urban hierarchy, but not necessarily in a dramatic way. Secondly, it reduces the lure of the gateway city between the constituent economies. Thirdly, it fosters some regional imbalance since one centre serves a growing market segment in the other region. Ultimately, this place becomes the only production centre for values of the fixed costs which are not very high.

These results lead us to expect that the construction of a high-speed connection will favour the regional centre which has comparative advantage. In other words, a stronger specialisation among mega–cities should emerge, at least for all the services using skilled and mobile workers as well as in sophisticated retailing activities. In addition, the lure of the primary regional centres should also be strengthened at the expense of more peripheral, secondary centres since the market area of firms located therein expands. These two observations agree with the analysis of the Japanese urban system made by Murayama (1994, p. 100) who observes that Shinkansen cities immediately gained the highest locational advantage, while non-Shinkansen cities in remote regions move further toward the periphery.

Despite some real appeal, our approach suffers from several limitations that should be accounted for in future research. Among the possible extensions that are worth studying, we would like to mention the following. Firstly, we restrict ourselves to one grid and one squared toy-network, although our approach can be extended to other configurations (more than two regions, other typical shapes of the transportation networks, other densities of demand points, other sizes of the lattices of points, other combinations of the gates or other the characteristics of the gates). Secondly, we have considered a static setting that could yield to the long–run location pattern. However, experiments should ideally be conducted in a dynamic environment in order to derive more realistic insights; dynamic price should be considered and included in existing dynamic modelling processes such as those developed by Nagurney and Zhang (1997a and 1997b). Thirdly, demand is here considered as uniform and independent of the location of the facilities and/or the infrastructure. Instead, one should allow for changing patterns of demand in interaction with facility and infrastructure locations. Lastly, one should develop a broader model dealing with several goods and services characterised by different cost parameters in order to study the impact of the transport network on the hierarchy of centres.

PART THREE

Location-allocation modelling and the measure of distance

Introduction to Part III

Space and distance are closely related: distance is the relation between the elements located in the space; it gives the opportunity for physical separation (see Beguin and Thisse, 1979; Huriot, Perreur and Derognat, 1994). Distance plays an important role in the fundamental explanation of spatial interdependence. It represents the friction introduced by the space in the exchanges (Perreur, 1989). This friction is not neutral. Spatial separation can be deplored or wanted. It is deplored when it hinders the realisation of some societal or individual person's objectives (for example it makes exchanges more expensive). It is wanted when, for instance, it prevents negative externalities; in this case, it protects by realising spatial segregation of economic activities or human groups. The same measure of spatial separation between two places can have positive effects for some agents and negative effects for others. Hence it is necessary to evaluate accurately and realistically the level of separation between two places, that is to say distance.

Applied location-allocation problems are often large in scale: a public agency may, for instance, wish to locate a number of facilities in order to meet the demand of the residents of an urban area. In such a case, every residence in the area is designated as a demand point with a corresponding weight proportional to its requirements; these points are separated by distances measured on a transportation network represented by a graph.

There is no unique representation of distance. Number of geographers and spatial analysts have already defined physical, time, economic or cognitive distances (Downs and Stea, 1977; Beguin and Thisse, 1979; Bailly, 1985; Muller, 1985; Perreur, 1989; Huriot and Perreur, 1992). Distance can be evaluated quantitatively or qualitatively, in terms of distance length (Beguin and Thisse, 1979), transport distance (Huriot and Thisse, 1985), time distance, distance itinerary (Huriot and Perreur, 1990), shortest-path distance (Huriot, Smith and Thisse, 1990), distance network, or more generally in terms of disutility. Distance can be observed or approximated. Distance can be objective or cognitive. It sometimes corresponds to a metric, sometimes not (Perreur, 1989; Huriot and Perreur, 1990). The evaluation of the distances enables to set the properties of the studied space. Geographers and economists are often forced to use simplified formal representations of space such as maps or to associate a mathematical space with the economic space.

With a few exceptions, spatial analysts use Euclidean geometry; this choice is not neutral as space is then Euclidean (Perreur, 1989).

Not all elements related to the 'distance measurement problem' will be tested in the following chapters. Three specific aspects will be analysed.

This research is concerned with the impact of the spatial morphology of the transportation network on the location models. Hence given a transportation structure, it is important to see how far the weights attributed to the links influence the location-allocation results. For these models, it is certainly ideal to have the highest possible level of detail in the input data. Hence myriads of travel distances need to be looked up. Using all the available data, however, may lead to intractability: there are too many demand points, too many distances to measure. Hence there is an advantage in using distance-predicting functions. When objects in space are represented by points, distance-predicting functions may be used to transform the point coordinate differences of two points into an estimate of the real travel distance between them. Hence, Chapter 6 uses l_p-metrics and tests how far the *k*-median results are sensitive to changes in the l_p-distance parameters. It will show that the *k*-median results are rather insensitive to the changes of the parameters of the l_p-distance.

Chapter 7 is concerned with another problem directly bound to transportation: the financing of the transportation cost. The results of the price location problem are compared for different price policies as well as different transportation networks. Using a price policy enables to take into account time, capital, work and the real cost of getting over the physical distance. How are distribution costs shared between firm and clients as well as among the clients? What are the consequences of such choices in optimal spatial organisation? Several price policies are used, showing the complex relations between geographical and economic distances. It is shown that the choice of a price policy has a limited impact on the firms' locations compared to the transportation configuration.

Many human activities generate effects on the neighbouring environment (such as pollution). The economic concept representing these effects is called 'externality'. The introduction of externalities in location-allocation models implies changes in the models' formulations as well as in the resolution methods: in addition to transportation costs, the model includes a minimisation–maximisation function of the externalities on the nearby environment. Both costs are unwanted by the users of the facilities: they want a facility as close as possible and they also want the negative effects they have to undergo as small as possible. Transportation costs are measured and produced on a transportation network; pollution costs are measured at nodes on the network but they diffuse all over the studied area. The issues of accessibility to facility services and minimising exposure to communities provide an interesting planning situation in that competing siting objectives are involved. Chapter 8 shows how far the location-allocation results are sensitive to the summation of such measures, to the trade-off between both objectives.

6. Distance predicting functions and location-allocation results

6. 1 INTRODUCTION

As previously mentioned, location-allocation models aim at determining the optimal number of facilities, the locations of the supply sites and which customers each site serves. Hence each application to an empirical case study requires a data-collecting step in order to represent the studied environment in a satisfactory way. In most real-world applications, the environment is represented by a discrete space where demand is located at nodes and distances are measured along links drawn between nodes. Many simplifications are made in order to represent this environment, leading to many well-known spatial data analysis problems: how much does the level of data aggregation affect location-allocation results? How much does the measure of demand influence location-allocation results? Does the shape of the road network affect the location-allocation results? Do boundaries affect location-allocation models? Data problems are numerous and many remain unsolved.

As in many spatial modelling processes, location-allocation applications lead to a variety of layouts for the data points that are virtually limitless: each empirical configuration is unique, with its own demand and supply points, shape of network and weight distribution. This often limits the degree of generality of the conclusions. This chapter isolates one data problem among others: it examines the effects of distance measurement on location-allocation results. More particularly, it considers the problem of testing the robustness of the solutions of a representative location-allocation model (the *k*-median) to the parameters defining a l_p-norm when the latter is used for estimating actual travel distances on transportation networks. Using a distance predicting function instead of real-world distance makes the data-collecting step easier. Hence a trade-off exists between computational tractability and model accuracy, and it is important that a sensitivity analysis is performed in conjunction with such a distance estimation procedure.

Measuring distances on a road network is a time–consuming task in most empirical spatial modelling processes. It involves the digitalisation of the network, the evaluation of the velocity along the different links, the computation of the shortest paths and the verification of often very large distance matrices. Hence this lengthy procedure is much error-prone (see Langevin, Mbagara and Campbell, 1996, pp. 165-6 for some examples). Hence, distance–predicting functions are used in order to transform coordinates differences between two points into an estimate of the travel distance between them (for example Brimberg and Love, 1993; Love and Morris, 1972, 1979 and 1988; Love, Walker and Tiku, 1995). Distance predicting functions rapidly provide estimates of unknown distances between pairs of points in a geographic region and can be easily implemented in geographical information systems. A commonly used function is the l_p-norm (see for example, Love and Morris, 1972, 1979, 1988; Brimberg and Love, 1992, 1993a and 1993b; Muller, 1982a). Fitting an l_p-norm to distances on a transportation network means finding empirical values of a parameter p as well as the orientation θ of the reference axes or also some weights (for example Brimberg and Love, 1992 and 1993). The empirical fits depend upon the studied example (see Sections 6.2.2 and 6.2.3 for more details) and are estimated by regression methods.

This type of estimation method enables an easy estimation of travel time and distance, including geographical or physical factors of each link in the route distance. It has been shown in environmental psychology that the distance perceived by consumers is affected by geographical or physical factors of the traversed route. For example the perceived distance is affected by the number of intersections and turns, change in landscape, information along a street and so forth (for example Sadalla and Staplin, 1980; Saddala and Magel, 1980; Allen, 1981).

As in most empirical fits, estimation errors are made on the empirical fit of p and θ: the unknown real-world distance between one pair of points may be over- or underestimated because of the statistical nature of distance–predicting functions. Our aim is not to discuss the confidence intervals for unknown distances (Love, Walker and Tiku, 1995), nor to find the best estimates for one particular case–study (Berens and Körling, 1985). The objective of this chapter is to test how much small changes in p or θ affect location-allocation results, or, in other words, does the precision for empirically estimating the parameters p and/or θ of a l_p-distance-predicting function really import for the robustness of k-median results?

The remainder of this chapter is organised into three sections. Section 6.2 outlines the formal terms of the location-allocation problem and the distance measurement problem; it describes the design of the experiments: the settlement system (Section 6.2.1), the distance–predicting function (Sections

6.2.2 and 6.2.3), how irregularities are introduced in the grid (Section 6.2.4) and the location model chosen (Section 6.2.5). In Section 6.3 we detail the effects of p (Section 6.3.1) and θ (Section 6.3.2) on the location results when $k = 3$, while we summarise the findings for irregular lattices of points (Section 6.3.3) and for other values of k (Section 6.3.4). In subsection 6.3.5, exponents are applied to distances. The final section (Section 6.4) gives some concluding comments. Most results presented in this chapter have been published in *Geographical Systems* (Peeters and Thomas, 2000) and *Les Cahiers Scientifiques du Transport* (Peeters and Thomas, 1997).

6.2 DESIGN OF THE EXPERIMENT

6.2.1 The Settlement System

As previously mentioned, choosing a regular lattice rather than an irregularly shaped layout enables one to isolate the tested problem from many other sources of variation and to control for the spatial layout (for example Anselin, 1986; Haining, 1986). The chosen hypothetical settlement system can be seen to differ from the shapes of many observed settlements encountered in socio-economic research. However it enables the researcher to conduct the tests without the additional complexities introduced by empirical lattices and leads to more clear-cut interpretations of the results.

The spatial layout of the data set is a regular 15 × 15 squared lattice of points ($N = 225$ points). The choice of the shape and size of the lattice is governed by the preceding results: it appears to be the best compromise between feasibility, reasonable computing time and practical results. Each point i is a demand point as well as potential supply site. Each point i is characterised by its (x_i, y_i) coordinates and is a node bound by a link to its eight closest neighbours. The way of measuring the length of the link is defined in Sections 6.2.2 and 6.2.3. Hence the studied squared network is composed of 196 elementary cells, similar to those illustrated in Figure 3.1.

Demand is equal for all nodes and normalised to one. Demand is totally satisfied by the k optimal locations of supply and is inelastic to distance, as everywhere in the simulations performed in this book.

6.2.2 The Distance–Predicting Function

When objects in space can be represented by points, distance-predicting functions may be used to transform the point coordinate differences of two points into an estimate of the travel distance between them. The use of distance–predicting functions for verifying the arc lengths between nodes in road network data was suggested by Ginsburgh and Hansen (1974). Because road network data often contains errors, it is advisable to screen the data before using it in any type of application. Although they can take different forms and be used in different circumstances, we used the l_p-norm, where p measures the rectangular bias of the road network (Love, Walker and Tiku, 1995).

The Minkowsky distance (or l_p-norm) is often used as a surrogate for travel time as it often provides a good fit to network distances (for example Love and Morris, 1972, 1979; Brimberg and Love, 1992, 1993; Huriot and Perreur, 1990). Its functional form is given by:

$$l_p(m,n) = (|d_x|^p + |d_y|^p)^{1/p} \tag{6.1}$$

where $d_x = |x_m - x_n|$

$d_y = |y_m - y_n|$

When $p < 1$, which is known as the 'hyperrectilinear' case (see Juel and Love, 1985), the triangle inequality postulate ($d(x,y) + d(y,x) \geq d(x,z)$) is not always satisfied. Hence l_p is not a norm in the topological sense. On the l_p-sphere, the circle bends inward (Figure 6.1 shows the locus of points whose distance with respect to a central point remains constant). A value of p smaller than 1 may however not be totally unrealistic in applied location studies: Muller (1985, p. 191) gives the example of a motorist taking a detour to avoid traffic as an evidence of a situation where the triangle inequality rule is no longer applicable. In this case, the corresponding space is a semi-metric (Perreur, 1989).

When $p = 1$, the distance is called rectangular, 'city-block', rectilinear, 'Manhattan' or l_1. The isovectors on the l_p-sphere are diamond-shaped: they take the shape of a square from which the diagonals are parallel to the axes (Figure 6.1). This distance yields in applications for cities where movements exist only along line segments parallel to two perpendicular axes (for example Wesolowsky and Love, 1971; Muller, 1985; Huriot and Perreur, 1990). When the street network is not orthogonal, the l_1-distance overestimates the real distance; this overestimation is increased when the system of axes is not well oriented. On the average, the rectilinear distance is 1.27 greater than the Euclidean distance (Huriot and Perreur, 1973, p. 209).

For every $p > 1$ and finite, the spheres are strictly convex. When $p = 2$, distance is called 'Euclidean'. It corresponds to the crow's flight; isovectors are circles on the l_p-sphere (Figure 6.1). It is the shortest distance between two points; Euclidean distance generally underestimates the distances on real networks; for short trips, the Euclidean distance gives a good approximation of the spherical distance (Huriot and Perreur, 1990; Perreur, 1989). If $p \cong 2$, the network is highly developed since travel ease is not significantly affected by the direction of travel (Morris and Love, 1979).

For values of $p > 2$, the distance is still a metric, but the corresponding Minkowsky distances are shorter than the Euclidean distance. Such distances are used by subjects in their ratings (see Muller, 1985, p. 191 for the relevance of such values of p). The balls in the limit case $p = \infty$ (dominance metric, sup metric or Tchebycheff metric) are squares whose sides are parallel to the axes; this metric is of no use in spatial representations (see Perreur, 1989). Let us also mention Lundberg (1973) who uses a p value of 3.0 as the best evaluation for subjective distances. The exponent p gives in fact the type of itinerary: crow's flight, rectilinear, and so on (Perreur, 1989).

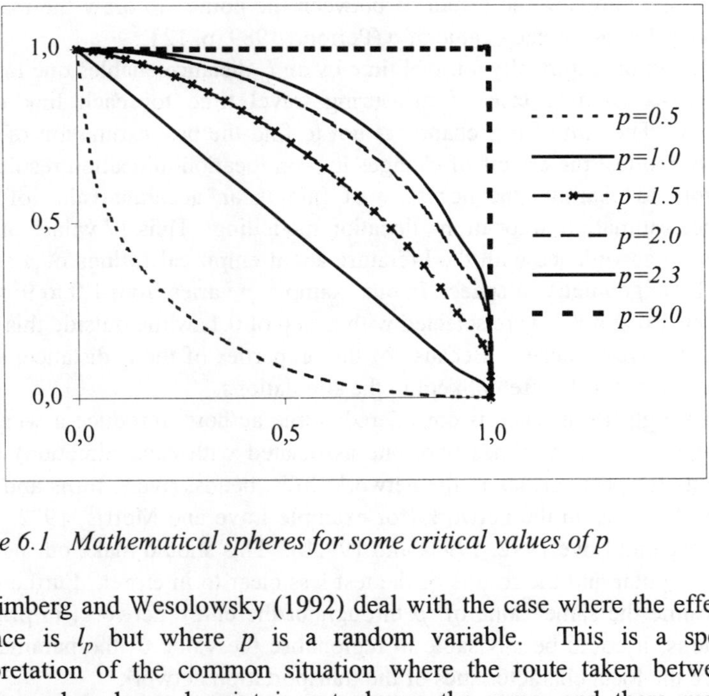

Figure 6.1 Mathematical spheres for some critical values of p

Brimberg and Wesolowsky (1992) deal with the case where the effective distance is l_p but where p is a random variable. This is a specific interpretation of the common situation where the route taken between a facility and a demand point is not always the same, and there may be probabilistic variations due to traffic conditions, road construction, driver unfamiliarity with best routes, or other factors.

The parameter p is usually determined by a numerical procedure minimising some goodness of fit criterion such as a weighted sum of squares deviations of absolute deviations. Values of p equal to 1.0 and 2.0 are often cited in applied transportation problems with no particular empirical justification (for example Langevin, Mbagara and Campbell, 1996).

Examples of empirical fittings of l_p-distances functions to actual data are however found in the literature. They aim at finding the best surrogate for travel time between cities or within cities. Empirical values of p vary from 0.9 to 2.29 (see for example Love and Morris, 1972, 1979; Muller, 1982b; Berens and Körling, 1985, 1988; Love, Truscott and Walker, 1985; Love, Morris and Wesolowsky, 1988; Brimberg and Love, 1993) and highly depend upon the studied regional environment. Most often computed measures of p vary between 1.5 and 2.0. Muller (1982b) also notes that road distances are not Euclidean. They are longer with characteristics of the rectilinear distances.

Note that these metrics can be used for measuring the distances on maps, knowing the coordinates of the points and fixing a value of p. Conversely, it is possible, knowing the distances between the points, to draw the map that will be different for each value of p (Perreur, 1989, p. 121).

Estimating empirically a travel time by an l_p-distance enables one to avoid the time-consuming task of measuring travel time for each link of the network. The aim of this chapter is not to find the best estimation of travel time, but to test the effects of changes in p on location-allocation results, and therefore to examine the necessity of finding an accurate value of p for distance estimation in location-allocation modelling. Thus 17 values of p are chosen in accordance with the literature about empirical fittings of p as well as with the geometry of space. In our example, p varies from 1.0 to 9: values between 1.0 and 2.3 were selected with a step of 0.1, while outside this range 3, 6 and 9 were chosen. Because of the properties of the l_p-distance, values of p smaller than 1 were not kept in the simulations.

No weight to distance is considered: some authors introduce a weight or inflation factor (sometimes two, one associated with each direction) which accounts for peculiarities in the network: hills, bends, rivers, turns and other forms of 'noise' in the network (for example Love and Morris, 1972, 1979; Brimberg and Love 1992, 1993a and 1993b). This should make our network more irregular and the results of the test less clear to interpret. Furthermore, we assume the same value of p throughout the entire network; in practical situations, it could be advisable to regionalise the value of the parameter to capture the local characteristics of the transportation network.

In the simulations proposed in this chapter, 17 measures of p in the l_p-distance are considered.

6.2.3 The Orientation of the Reference Axes

More accuracy in estimating travel time by an l_p-distance can be obtained by allowing a rotation of the orthogonal reference axes (see for example Huriot and Perreur, 1973; Love and Morris, 1973; Brimberg and Love, 1992, 1993a and 1993b; Love, Morris and Wesolowsky, 1988). θ is the rotation parameter, the angle measuring the rotation of the orthogonal reference axes X and Y, the original reference axes being those parallel to the edges of the lattice of points. θ is measured counter–clockwise; it measures the relative difficulty of travel in any direction: if $p < 2$, it is easier to travel along the directions of the axes, while if $p > 2$, it is easier to follow the direction of the bisectors. When $p = 2$, space is isotropic: the measure of distance does not depend upon the direction or orientation of the axes (for example Love and Morris, 1979; Brimberg and Love, 1993a). Brimberg and Love (1992, p. 131) show that if $r(\theta)$ is called the directional bias function of the norm k, in a qualitative sense, this function can be thought of as a measure of the relative difficulty of travel in any direction. If $r(\theta_1) > r(\theta_0)$, then one must travel a longer distance along a line at angle θ_1 with the x_1-axis than along a line at angle θ_0 with the x_1-axis to cover the same Euclidean distance between pairs of points.

Hence, Brimberg and Love (1993) have shown that the search for the best-fitting value of p can be restricted to the interval [1,2] (or more generally [0,2] when hyperrectilinear distances are included) provided the reference axes are oriented to match the directional bias of the distance function with that of the network.

In accordance to the demonstrated symmetry about $\Pi/4$ (Brimberg and Love, 1992), θ takes the values 0°, 15°, 30° and 45° in our simulations. These rotations are applied for all p values. For each of the 17 values of p, four values of θ in the l_p-distance predicting function are considered, leading to a total number of 68 simulations.

6.2.4 Irregular Lattices

The choice of a regular lattice of points may be questioned on grounds that it yields too–specific results. To assert the robustness of our findings on irregular distributions, we have proceeded in the following way. Each of the 225 points of our regular lattice has been considered as the mean of a bivariate normal distribution characterised by the same standard error σ in the x and y directions. For a given value of σ and each of the 225 distributions, a single point has been created using a pseudo-normal generator. The selected values for σ, which can be viewed as a measure of disorder of the lattice, were 0.00 (regular), 0.25, 0.50 and 1.00. On the four lattices obtained in this way, selected values of p and θ detailed in Section 6.3.3 were used to measure the distances.

Experiments were also carried out using a set of 225 points randomly generated in a square. This can be considered as a maximum disordered set.

6.2.5 The Location Model

The k-median model is applied (Section 3.2). It yields a configuration of k supply sites such that the sum of the distances between the demand points and their respective closest supply site is minimised. The model has been chosen for its high flexibility, the good performance of the algorithms designed to solve it, and its common use in location-allocation applications. Most importantly for the purpose of this chapter, distance is one of the key elements in the formulation of the k-median problem: it is the only criterion optimised in the objective function. Distance is often used as a surrogate for travel disutility encountered by the customers; thus using a 'good' measure of distance is one of the crucial conditions for obtaining meaningful solutions. As computing actual distances between points in real spaces is generally time-consuming and prone to errors, approximating those distances by analytical expressions using the coordinates of the points, as geographical information systems allow, seems a priori an interesting idea. Exact solutions of the k-median model were computed using a program outlined in Hanjoul and Peeters (1985).

The optimal configuration of supply sites and the value of the total distance are not the only characteristics of a solution of a k-median problem. The number of demand points allocated to a site gives the level of activity of this site, thus its size in terms of human and material resources that should be affected. The shape of the area to be supplied by a site and the corresponding distribution of distances between the demand points and the site are two other important outcomes. The latter may be characterised by its mean value, its maximum value, its variance, a Lorenz curve and so on. All the measures of spread can be interpreted in terms of inequity among clients.

In the studied example, the number of supply sites k to be optimally located may theoretically vary between 1 and 225. In this chapter, we decided to focus on the results obtained for $k = 3$; according to our experience, it was the value that best isolates the tested effect as it leads to a symmetric solution in the chosen lattice of points. However, to assess the robustness of the findings, experiments with other values of k were also performed. The results are synthesised in Section 6.3.4.

6.3 COMPUTATIONAL EXPERIENCE

6.3.1 How much does the Choice of p Affect the k-median Results?

Let us consider a first set of simulations: a regular squared lattice of 225 points, reference axes parallel to the lattice of points ($\theta = 0$) and p varying from 1 to 9. Let us consider the outputs of the k-median for the 17 selected values of p.

Average and maximum distances

Two measures of distance commonly help to evaluate the k-median results: (1) the average distance (D_{ave}) between each demand point and its closest optimal supply site gives a measure of efficiency of the proposed solution, and (2) the maximum distance (D_{max}) between a demand point i and the closest facility can be considered as a measure of the equity of the optimal solution. Their variations with p are respectively reported in Figures 6.2 and 6.3.

A *3*-median model was applied for each of the 17 values of p. Spatial differences between the solutions were found because of the isotropy of the studied sample of points: symmetric solutions were suggested by the model. In this particular case, the three optimal sites are located at the vertices of a triangle, sometimes on its base, sometimes on its summit, sometimes rotated by 45° or 90°. Hence comparability of the optimal locations was difficult for geometric reasons only: for a given value of the objective function, different solutions can be suggested (alternative optima); this is simply due to the high degree of symmetry in the studied square. A second set of simulations was therefore performed, imposing one location to the model. This enables one to avoid natural symmetries, to better compare the spatial solutions and to interpret the tested effects without confusion. In this studied example, imposing one location has no effect on the average distances and little on the maximum distance for values of $p < 1.5$ (Figure 6.3). The following sections will refer to the constrained model only.

Both distances (D_{ave}, D_{max}) (Figures 6.2 and 6.3) decrease with p, in accordance to the definition of the l_p-distance (Section 6.2). Figure 6.2 compares the relative variation with p of the average distance (D_{ave}), the maximum distance (D_{max}) as well as the expected distance (distance simply computed with the l_p formula). Differences are slight and insignificant. The ratio of the maximum distance (D_{max}) to the average distance (D_{ave}) is often considered as a balance between equity and efficiency. In our simulations, this ratio varies regularly with p, but the fluctuation is very slight: between 2.15 to 2.26 with a minimum value when $p \in [1.8;3.0]$. Relative differences in the computed values of D_{ave} and D_{max} are not insignificant with p, but regular.

p influences the absolute value of the average and the maximum distances; the observed variations with p are proportional to the expected one (Figure 6.2).

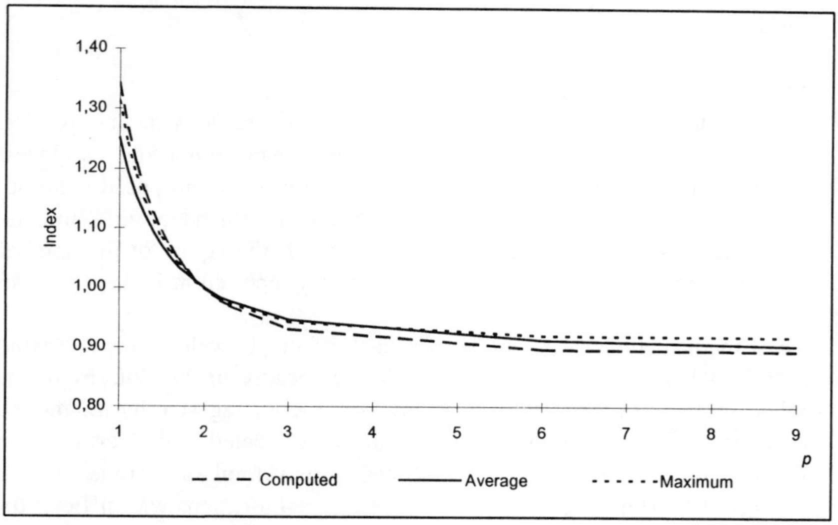

Figure 6.2 *Relative variation with p of the computed l_p-distance and 2 outputs of the 3-median model: the average distance and the maximum distance (Ratio = $X_p/X_{p=2}$)*

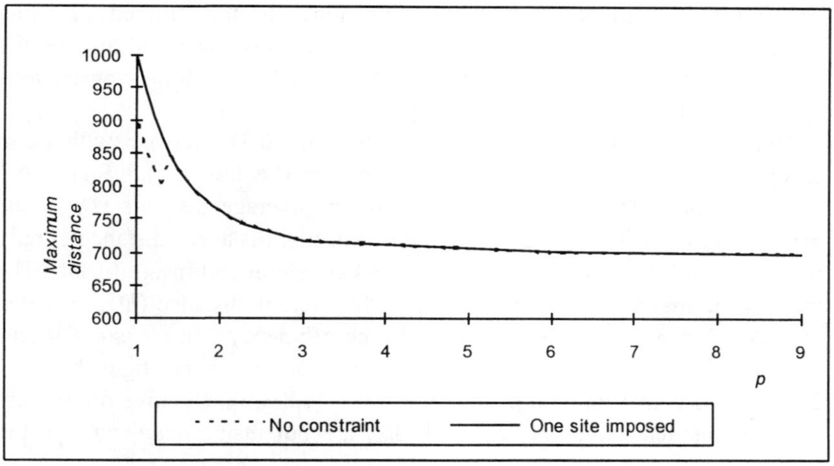

Figure 6.3 *Computed variation of D_{max} with p for the 3-median model applied with no constraint and with one imposed location*

Optimal locations and allocation

For $k = 3$ and one site imposed ($j = 53$), the same two other locations are always retained as optimal, for every p in [1,9] (Figure 6.4): this suggests that for this set of experiments, the optimal locations are not sensitive to the choice of the l_p-distance.

Each demand point i is allocated to its closest optimal supply site j. Each supply point j is therefore characterised by a zone (a market or service area) including the allocated demand points, and by the size of the facility, which is proportional to the number of points in the supply area. Both criteria (quantity of demand allocated to the site; shape of the zones) are used to evaluate the k-median results. In accordance with the preceding results, the design of the optimal market areas are almost the same whatever p in [1,9]. Figure 6.4 gives the three optimal supply sites j as well as the 11 demand points i whose allocation changes with p. The observed differences are located along the perpendicular bisectors of the triangle formed by the three optimal locations and are observed for very large values of p ($p > 3.0$). These locations are explained by the geometry of the studied environment and can easily be simulated by Voronoi diagrams (for example Sakakibara, Matsuyama and Suzuki, 1996).

Hence errors made on the use of p when estimating distance by the l_p-distance have little influence on the optimal locations and allocations proposed by the k-median model. No difference is observed for commonly used values of p (Section 6.2.2).

Distribution of distances

Let us make a last verification of the tested hypothesis. The distribution of the distances between each demand point i and its closest optimal supply site j is another way of evaluating the efficiency and the equity of the optimal solutions of location-allocation models. Figures 6.5 to 6.7 illustrate the variation of three simple statistics that describe the statistical distributions of a variable: (1) the coefficient of variation (CV) which measures the dispersion of a variable independently of the size of the data values; CV gives the ratio of the standard deviation against the mean ($CV = s/m$, sometimes multiplied by 100); (2) the coefficient of skewness (sk) takes the value 0 when the data are perfectly symmetric, it is positive when the mean is larger than the median and negative when the mean is smaller than the median. sk measures the tendency of a distribution to stretch out in a particular direction; (3) the kurtosis (K) measures the peakedness of the distribution; K is small if the frequency of observations close to the mean is high and the frequency of observations far from the mean is low.

Whatever the studied indicator, variations with p are very slight and insignificant; they are, however, regular with p: sk and K increase regularly with p, CV reaches minimum values when $p \in [1.5;2.1]$.

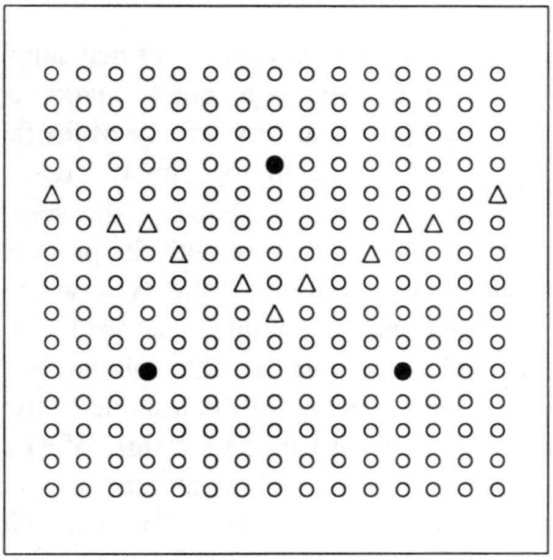

• = optimal location ○ = demand point Δ = changes in allocation

Figure 6.4 Optimal locations and allocation changes with p, when $\theta = 0°$

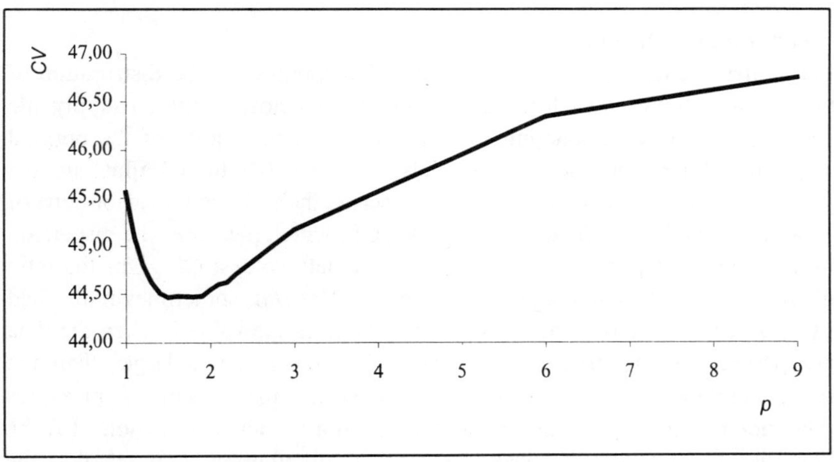

Figure 6.5 Variations of the variation coefficient (CV) measured on the distribution of distances with p when $\theta = 0°$

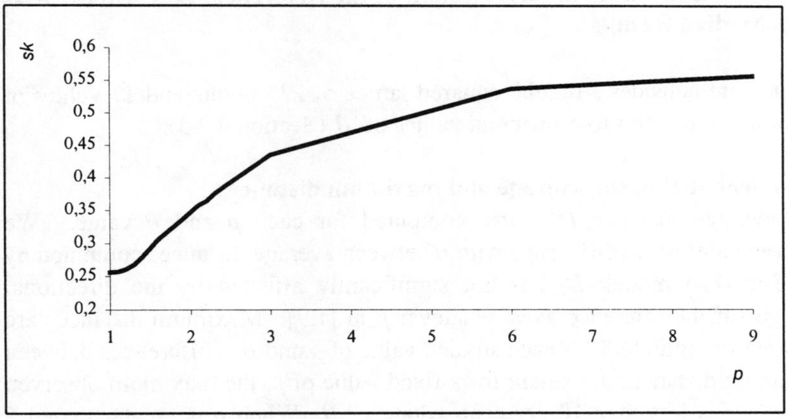

Figure 6.6 Variations of the skewness coefficient (sk) with p when $\theta = 0°$

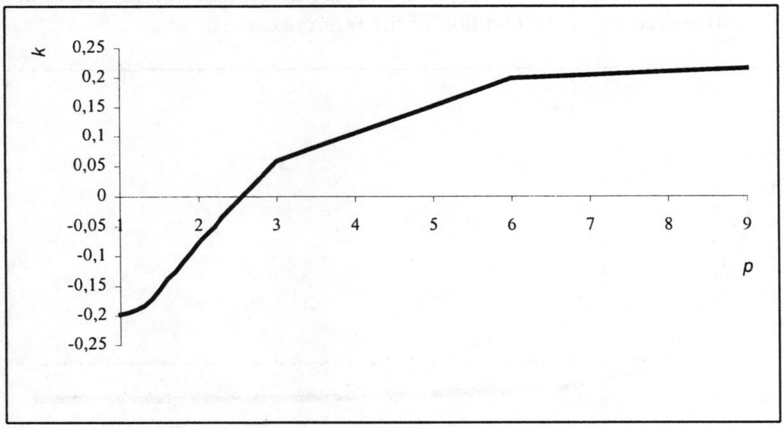

Figure 6.7 Variations with p of K measured on the distribution of distances when $\theta = 0°$

Conclusions

Within the limits of these experiments, it appears that p has little influence on the location-allocation results, especially in the range widely used in practice ($p \in [1.5\text{-}2.0]$). Optimal locations do not vary with p. Optimal allocations change only slightly at the borders of the market areas. The absolute values of the average and maximum distances vary only as theoretically expected by the distance formulation.

6.3.2 How much does the Orientation of the Reference Axes Affect the *k*-Median Results?

Let us here consider a regular squared lattice of 225 points and 17 values of *p*, but this time with four different values of θ (Section 6.2.3).

The effect of θ on the average and maximum distances
The average distance D_{ave} was computed for each *p* and θ value. We observed almost no difference with θ between average distances computed by the *3*-median model: D_{ave} is not significantly affected by the directional changes of the reference axes, whatever *p* in [1,9]. Maximum distances are reported in Figure 6.8 for each studied value of *p* and θ. Differences between maximum distances are small: for a fixed value of *p*, the maximum observed difference is less than 10 per cent, whatever θ. When *p* is greater than 1.5 and smaller than 2.3, differences are the smallest: they are less than 3 per cent.

As a first conclusion, we can say that average and maximum distances are slightly influenced by the orientation of the reference axes.

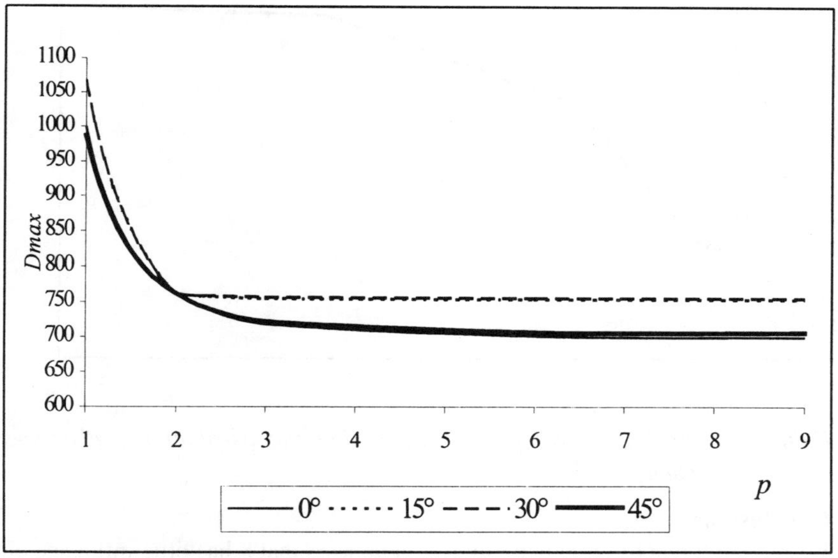

Figure 6.8 Variation with p and θ of the maximum distance (D_{max})

Figure 6.9 compares the maximum distance to the average distance for each value of *p* and θ. The computed ratio expresses the balance between equity and efficiency: the higher the index, the more equity is privileged against efficiency. Differences are not important: the index varies between 2.1 and 2.4. When $p > 2.0$, $\theta = 0°$ solutions are always the best balance between equity and efficiency. When $p < 2.0$, $\theta = 45°$ is the most balanced solution; this is to be explained by the symmetry demonstrated by Brimberg and Love in 1992 and 1993b. This figure also illustrates the equivalence of the solutions when Euclidean distance is considered (p = 2) and the resemblance between $\theta = 30°$ and $\theta = 15°$ solutions. Differences are small and regular with p.

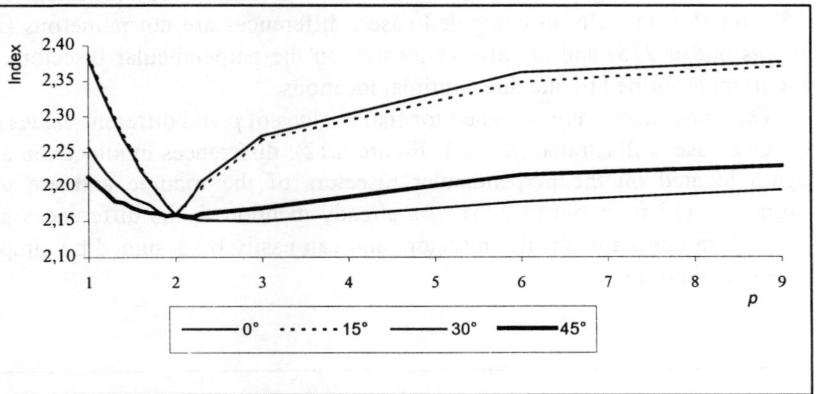

Figure 6.9 Variation with p and θ of the ratio between maximum (D_{max}) and average distances (D_{ave})

As a conclusion, the two most often used 3-median operational outputs are almost not affected by the direction of the reference axes as far as the commonly used values of p in the l_p-distance are concerned ($p \in [1.5 - 2.3]$). Errors on the use of p when estimating travel time (Section 6.4.2) do almost not affect the 3-median results as far as D_{ave} and D_{max} are concerned. Operational location results (choice of j, allocation of i to j) are the same whatever p and θ.

The effect of θ on the optimal locations and allocations

One site is imposed on to the model (j = 53); the other two optimal sites proposed by the 3-median are almost always the same as those illustrated in Figure 6.4. In one case out of 68, a shift of one site is observed; it occurs for a value of p = 1. In all other cases, whatever θ and p, the same supply sites are chosen and hence the solutions are stable.

Figure 6.10 shows the variation with p and θ of the quantities of demand allocated to the three optimal supply sites. Small differences between minimum and maximum allocation correspond to a balance between supply sites; large differences should not be chosen by the decision-maker especially when the service is not hierarchically organised. The figure shows that all p and θ values naturally lead to almost the same balanced configuration: p and θ have little influence on the allocation results (difference smaller than 10 per cent). For geometry reasons only, values of p close to 2 lead to totally identical values of maximum allocations.

Figure 6.11 illustrates the demand points i whose allocation varies with p for a fixed value of θ, that is to say the main differences in the shape of the market areas for a given set of reference axes. Two cases are illustrated: $\theta = 15°$ and $\theta = 45°$. In all computed cases, differences are not numerous (17 points out of 225) and are always located on the perpendicular bisectors of the triangle formed by the three optimal locations.

The same maps were designed for fixed values of p and different values of θ. One case is illustrated ($p = 1.1$, Figure 6.12): differences in allocation are again located on the perpendicular bisectors of the triangle and are not numerous (13 pairs out of 225). As already mentioned, the differences are located on the perpendicular bisectors and can easily be designed by simple Voronoi diagrams.

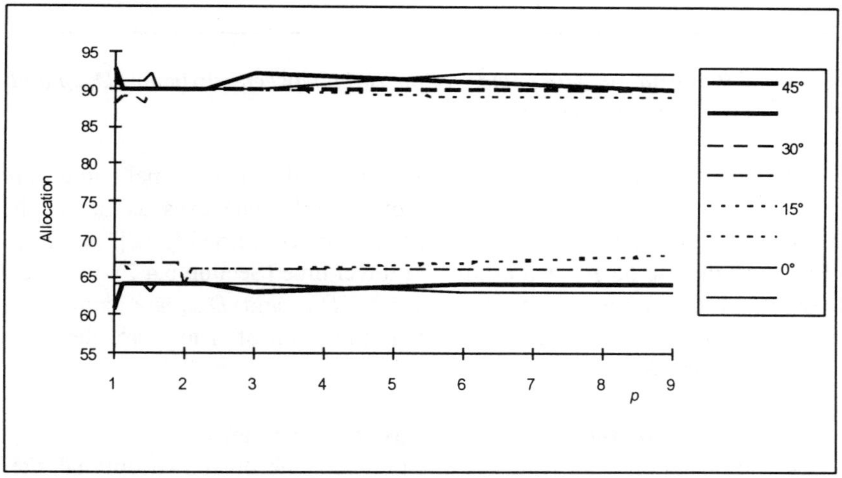

Figure 6.10 Variation with p and θ of the minimum and maximum allocation

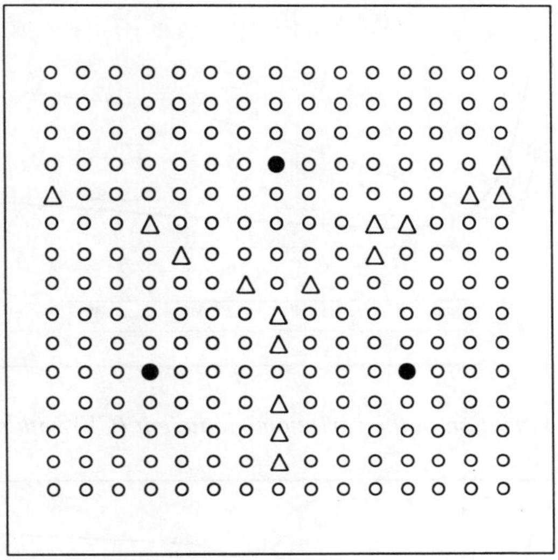

Figure 6.11 Optimal location (•) and allocation (Δ) changes with p (θ = 15° and 45)

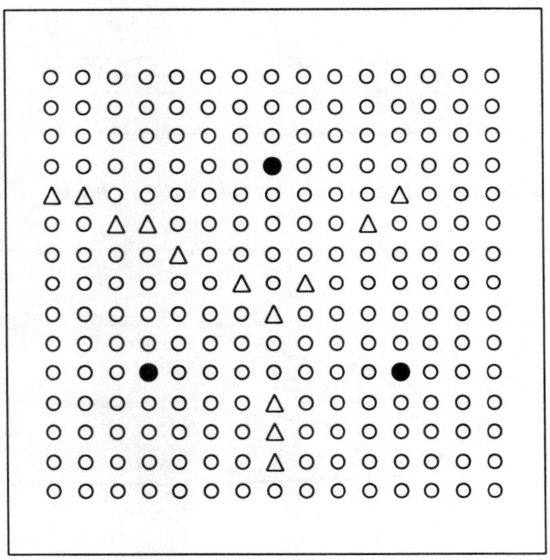

Figure 6. 12 Optimal location (•) and allocation (Δ) changes with θ, when $p = 1.1$

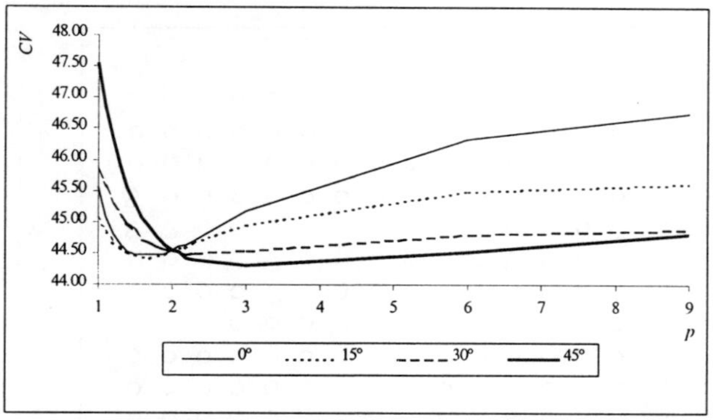

Figure 6.13 Variation of the variation coefficient (CV) with p and θ

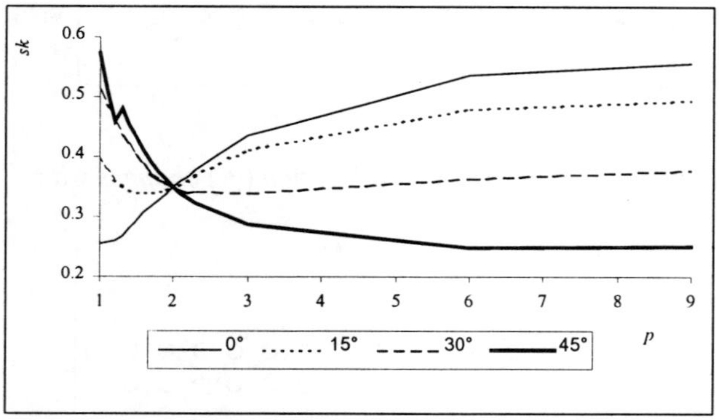

Figure 6.14 Variation of the skewness (sk) with p and θ

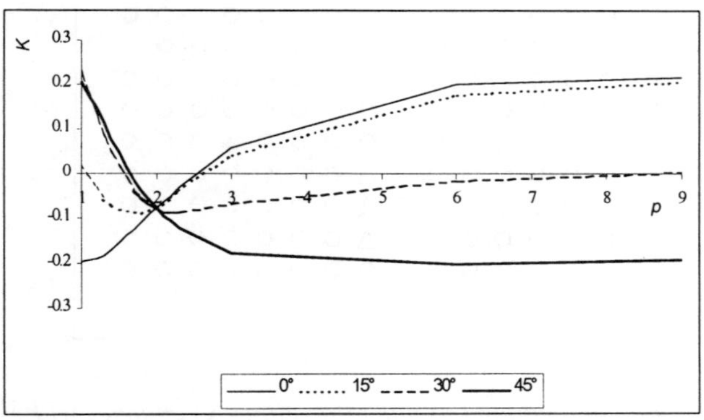

Figure 6.15 Variation of the kurtosis (K) with p and θ

The effect of θ on the distributions of distances

As in Section 6.3.1, the distributions of distances are summarised by three indicators: *CV*, *sk* and *K*. Figures 6.13, 6.14 and 6.15 refer to their variations with p and θ. Once again, differences are slight and not significant, but regular with p. Euclidean distances are not sensitive to θ changes. For geometric reasons only, solutions with $\theta = 0°$ are symmetric to those with $\theta = 45°$.

Evaluation of the relative effects of θ and p

Tables 6.1 and 6.2 summarise the amplitude of the effects of the changes in p and θ on four previously measured indicators: the average and maximum distances, the ratio between equity and efficiency, and the coefficient of variation measured on the individual distances between each demand point and its closest supply site.

Table 6.1 gives the relative changes of the four selected indicators with p. For each value of θ, the relative difference of each indicator is computed. Differences on the absolute values of D_{ave} and D_{max} are important due to the distortion of the space ($p \in [1; 9]$). Computing the differences for the ratio of D_{ave} to D_{max} lead to differences smaller than 11 per cent. *CV* is quite stable: *CV* is a relative value and has a maximum difference of 7.3 per cent between solutions. Relative differences with p between computed indicators are always smaller than 10 per cent.

Table 6.2 gives the relative changes with θ for four fixed values of p. Changes are slight whatever the studied indicator (smaller than 10 per cent).

Whatever p in [1,9] and θ, differences in location-allocation results are almost always less than 10 per cent, that is to say small in comparison to many other sources of errors that can be introduced when measuring real-world inputs.

Table 6.1 Maximum differences between the computed D_{ave}, D_{max} and CV for some fixed values of θ

θ	D_{max}	D_{ave}	*Ratio*	*CV*
0°	42.9	38.9	5.0	5.1
15°	41.5	40.6	10.6	2.7
30°	40.9	40.6	10.4	3.0
45°	39.6	40.7	3.5	7.3

Note: index = 100* $(X_{max} - X_{min})/X_{min}$

Table 6.2 Maximum differences between observed values for some fixed values of p

p	D_{max}	D_{ave}	*Ratio*	*CV*
1.0	8.1	0.9	7.8	5.6
1.5	3.6	0.3	3.9	1.9
2.2	1.6	0.0	1.6	0.5
9.0	8.0	0.6	8.3	4.3

Note: index = 100* $(X_{max} - X_{min})/X_{min}$

6.3.3 How Far does Irregularity in the Lattice of Points Affect the *k*-median Results?

In order to evaluate the robustness of the conclusions discussed in the preceding sections, further simulations were performed on irregular lattices of points. Two sets of experiments were conducted.

The first set of simulations refers to an irregular lattice of 225 points generated randomly (uniform random selection of the x and y coordinates; see Figure 6.18). We want to see if the results obtained in Section 6.3.1 about the changes of p on the *k*-median results still hold when the lattice of points is irregular. A *3*-median model is run successively with values of p varying from 0.4 to 9.0. As in the previous experiments, demand is also set equal to 1 in each node. Results are presented in Figures 6.16 to 6.18.

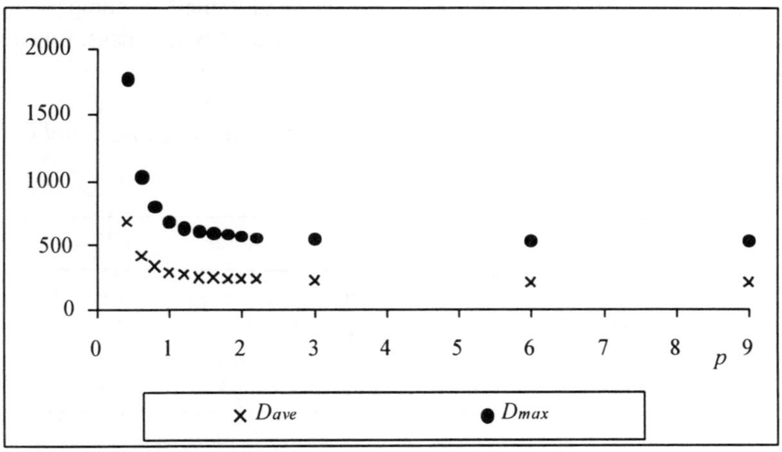

Figure 6.16 Variation with p of the average and maximum distances for an irregular lattice

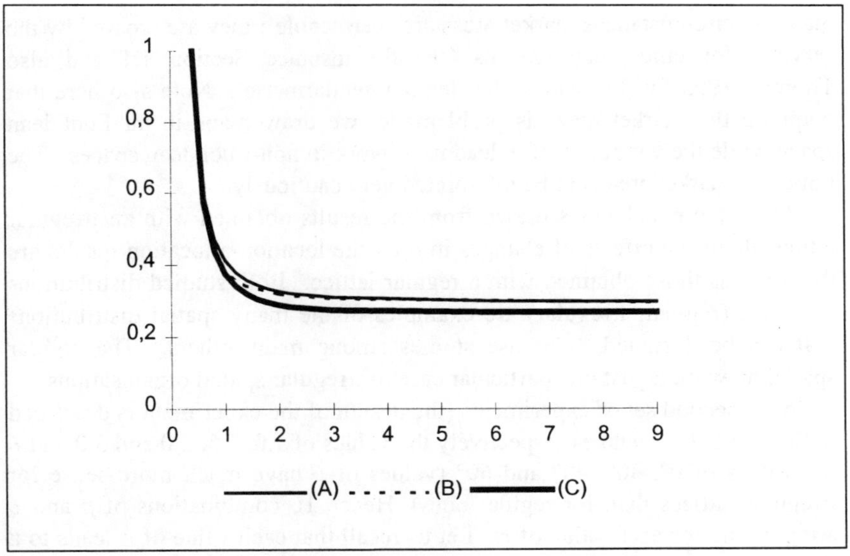

Figure 6.17 Variation with p of the average (A), maximal (B) and expected (C) distances

As mentioned in the case of the regular lattice of points, both average and maximum distances decrease with *p*, in accordance to the definition of the l_p-distance (Figure 6.16). Figure 6.17 compares the relative variation with *p* of the average distance, the maximum distance and the expected distance (distance simply computed with the l_p formula). Differences between the three curves are, once again, slight and insignificant. Relative differences in the computed values of the average and maximum distances are slight with *p* and regular. Table 6.3 gives the index of the three optimal sites selected for each value of *p*: *p* affects the choice of the optimal sites, but for empirically sustainable values of *p* (between 1.5 and 2.3) no change is observed in the choice of the sites. In other words, the only values of *p* that really affect the optimal locations are those empirically not sustainable (values lower than 1 or greater than 3, see Section 6.2).

Figure 6.18 shows the observed changes in the optimal locations on a map: they are slight. Allocation is discussed in Table 6.3: it gives the quantity of demand (here expressed in number of points) allocated to each site; these quantities do not change for values of *p* close to 2; they change erratically when *p* takes a value between 1.0 and 1.6 and change more importantly when *p* is smaller than 1 or greater than 3. The maps of the market areas are not represented here; the differences are slight (they are located at the limits of the designed market areas) and will not influence decision-makers as in many planning circumstances market areas are 'permeable': they are crossed by the services for emergency reasons (see for instance Section 1.1 and also Thomas, 1993 for the case of the Belgian gendarmerie). Note also here that mapping the market areas is problematic: we draw maps in an Euclidean space while the variations of p lead us to work in non-Euclidean spaces. The notion of market area is to be interpreted very cautiously.

Hence the conclusions drawn from the results obtained with an irregular lattice about the effects of changes in p on the location-allocation results are the same as those obtained with a regular lattice. Both studied distributions of points (regular, irregular) are examples of the many spatial distributions that can be designed, two case–studies among many others. The regular spatial network is just one particular case of irregular spatial organisations.

In the second set of experiments (the design of the experiment is described in Section 6.2.4), p takes respectively the values of 1.0, 1.5, 2.0 and 3.0 and θ the values of 0°, 30°, 45° and 60° (values of θ have much more sense for irregular lattices than for regular ones). Hence 16 combinations of p and θ are possible for each value of σ. Let us recall that each value of σ leads to a different lattice of points, the value 0 meaning perfect regularity of the distribution. For a 3-median model (with one location being fixed in $j = 53$), each value of σ leads to a slightly different spatial solution (local optima) but there is little or no change in the optimal locations with p and θ for a given σ (Table 6.4). Once again, the results put forward in the preceding sections are confirmed.

Table 6.3 Variation with p of the selected optimal sites (j1, j2, j3) and their allocated demand (A1, A2, A3).

p	*j1*	*j2*	*j3*	*A1*	*A2*	*A3*
0.4	41	165	143	85	65	78
0.6	41	171	143	83	64	78
0.8	41	150	165	86	80	59
1.0	41	143	166	83	73	69
1.2	47	143	166	84	74	67
1.4	47	143	166	82	75	68
1.6	47	143	166	85	74	66
1.8	47	143	166	83	75	67
2.0	47	143	166	83	75	67
2.2	47	143	166	83	75	67
3.0	47	143	166	82	77	66
6.0	56	129	166	74	89	62
9.0	56	129	166	74	89	62

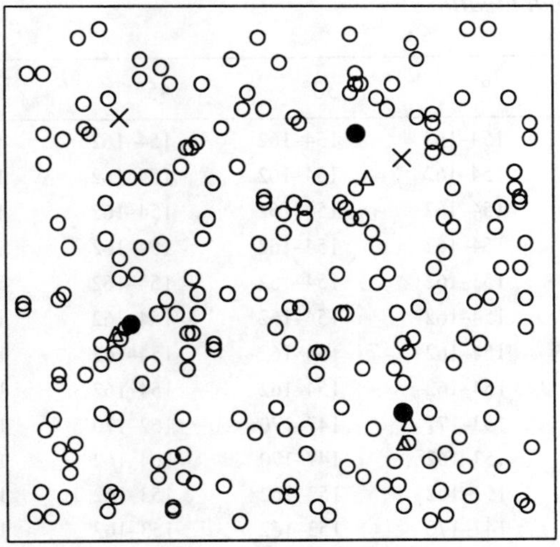

Note: O = demand point, • = frequent optimal location, Δ = not frequently selected site when $p < 1.0$, × = not frequently selected site when $p > 3.0$

Figure 6.18 Optimal location in an irregular lattice.

All experiments lead to the same conclusion: when we estimate a time-distance between two nodes of a network by a $l_{p,\theta}$-distance, the value of p (and θ) has little impact on the operational results of the k-median. The observed differences depend mainly upon p: (1) when p is very small ($p < 1$) differences are very important because distance does not respect the triangular inequality any more; (2) when p takes a value between 1.0 and 1.6, differences are mainly observed in the optimal allocations, not in the locations or the decision criteria (for example average and maximum distances); (3) when p is close to 2, no difference in the results are observed (Euclidean distance); (4) when p is large (greater than 2.2) important differences are again observed in the location-allocation results: computed distances are shorter than the Euclidean distance (Section 6.2). Let us here recall that empirically justified distances vary between 1.5 and 2.0 (see Berens and Körling, 1985), that is to say values that have – in the results of our experiments – little or no effect on the optimal results. These results are yielded of course only if the same l_p-distance is applied to all edges of the network, that is to say if the planner adopts the same predicting function for the entire study area.

Table 6.4 Variation with p, σ and θ of the 3-median solutions (index of the 2 optimal locations)

σ	p	0°	30°	45°	60°
	1.0	154–162	154–162	154–162	154–162
0.00	1.5	154–162	154–162	154–162	154–162
	2.0	154–162	154–162	154–162	154–162
	3.0	154–162	154–162	154–162	154–162
	1.0	153–162	154–162	154–162	154–162
0.25	1.5	154–162	154–162	154–162	154–162
	2.0	154–162	154–162	154–162	154–162
	3.0	154–162	154–162	154–162	154–162
	1.0	153–177	147–170	162–170	138–162
0.50	1.5	153–162	147–170	164–162	138–162
	2.0	153–162	153–162	153–162	153–162
	3.0	147–170	153–162	153–162	153–162
	1.0	91–176	138–162	110–162	138–162
1.00	1.5	138–162	138–162	138–162	138–162
	2.0	138–162	138–162	138–162	138–162
	3.0	138–162	138–162	138–162	138–162

6.3.4 Do Other Values of *k* Yield Significantly Different Results?

The same experiments were conducted for other values of k (number of facilities). They lead to the same conclusions, but are sometimes less clear-cut to present because of the presence of numerous alternate optima. Let us take the example of $k = 9$ (Table 6.5).

For a regular lattice of 225 points ($\sigma = 0$), very slight differences are observed if none of the nine sites is imposed. When imposing two locations (in $j = 38$ and $j = 113$) in order to avoid alternate optima, the optimum location patterns are totally identical, whatever p and θ. With other values of σ (0.25, 0.50 or 1.00), slight differences appear only for values of $p = 1$, all other solutions being equal, whatever p and θ. Hence we can here confirm the results obtained and discussed for $k = 3$.

Table 6.5 Variation with p, σ and θ of the 9-median solutions (index of the 7 optimal locations, 2 others being constrained (38, 113))

σ	p	0°	30°	45°	60°
0.00	1.0	33-43-108-118-183-188-193	33-43-108-118-183-188-193	33-43-108-118-183-188-193	33-43-108-118-183-188-193
	1.5	33-43-108-118-183-188-193	33-43-108-118-183-188-193	33-43-108-118-183-188-193	33-43-108-118-183-188-193
	2.0	33-43-108-118-183-188-193	33-43-108-118-183-188-193	33-43-108-118-183-188-193	33-43-108-118-183-188-193
	3.0	33-43-108-118-183-188-193	33-43-108-118-183-188-193	33-43-108-118-183-188-193	33-43-108-118-183-188-193
0.25	1.0	18-58-93-149-171-182-206	33-43-108-118-183-188-193	33-43-108-118-183-188-193	33-43-108-118-183-188-193
	1.5	33-43-108-118-183-188-193	33-43-108-118-183-188-193	33-43-108-118-183-188-193	33-43-108-118-183-188-193
	2.0	33-43-108-118-183-188-193	33-43-108-118-183-188-193	33-43-108-118-183-188-193	33-43-108-118-183-188-193
	3.0	33-43-108-118-183-188-193	33-43-108-118-183-188-193	33-43-108-118-183-188-193	33-43-108-118-183-188-193
0.50	1.0	32-42-109-137-171-181-206	42-47-119-139-188-193-198	28-38-108-119-182-188-193	32-44-108-148-182-188-208
	1.5	32-42-108-119-182-188-193	32-42-108-119-182-188-193	32-42-108-119-182-188-193	32-42-108-119-182-188-193
	2.0	32-42-108-119-182-188-193	32-42-108-119-182-188-193	32-42-108-119-182-188-193	32-42-108-119-182-188-193
	3.0	32-42-108-119-182-188-193	32-42-108-119-182-188-193	32-42-108-119-182-188-193	32-42-108-119-182-188-193
1.00	1.0	32-42-109-134-171-181-206	32-47-119-139-188-193-198	28-32-108-119-182-188-193	32-44-108-148-182-188-208
	1.5	32-42-108-119-182-188-193	32-42-108-119-182-188-193	32-42-108-119-182-188-193	32-42-108-119-182-188-193
	2.0	32-42-108-119-182-188-193	32-42-108-119-182-188-193	32-42-108-119-182-188-193	32-42-108-119-182-188-193
	3.0	32-42-108-119-182-188-193	32-42-108-119-182-188-193	32-42-108-119-182-188-193	32-42-108-119-182-188-193

6.3.5 Some Further Tests with Another Transformation of Distance

Our aim is not to extend the experiment to all possible distance-predicting functions. Preceding sections have shown that small changes in p and θ in the $l_{p,\theta}$-distance predicting function affect only very slightly the k-median results. Hence further simulations were performed with another commonly used transformation of distance: the power transformation (d_{ij}^{α}).

The exponent α expresses the friction of distance. This enables us to introduce equity in the k-median model when α is greater than 1. Richard, Beguin and Peeters (1990) indeed suggested introducing an exponent to distance as a compromise between efficiency (k-median) and equity (p-centre) in a problem dealing with the optimal location of fire stations within a

Belgian regional context. Empirically, they observed a rather good stability of the *k*-median solutions with respect to the introduction of equity; in their example, α varies between 1.0 and 1.4. When α is greater than 1, it expresses the propensity for equity; values smaller that 1 were also selected in our example in order to express the often used transportation functions. In our case, the exponent α varies systematically between 0.3 and 1.4 with a path of 0.1. Hence 11 values of α are used. Given the preceding results, there was no need to present optimal location results for all combinations of p and α values. Results are presented here for l_p-distances where p =1.7 and θ= 0° and for a squared regular lattice of points (Figure 6.19).

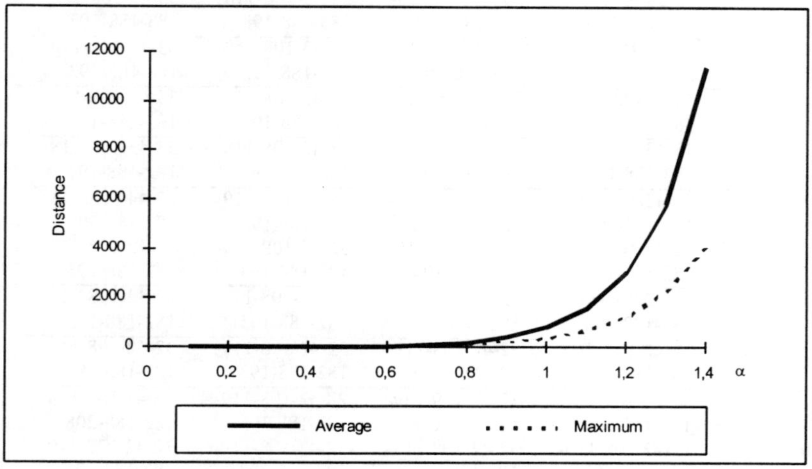

Figure 6.19 Variation with α of the average and maximum distances

There is no doubt about interpreting the location-allocation results: all simulations lead to the same three optimal locations, whatever the exponent applied to distance. α does not influence the choice of the optimal sites, nor the quantity allocated to them. Interpretation of the distances is clear: the absolute values of the average and the maximum distances increase exponentially, especially for values of α greater than 1 (Figure 6.19). When α is greater than 1, computed distances become much larger, due only to the definition of distance. However, these changes do not imply any changes in the optimal locations.

When real-world distance is transformed by a power function and when this transformation has the same amplitude for all segments of the studied transportation network, then changes are only observed in the evaluation of the model parameters (average or maximum values) but no change is observed in the optimal location results.

6.4 CONCLUSIONS

The objective of this chapter was to simply test how robust the solutions of a location-allocation model are to changes in the parameters of a distance-predicting function and consequently with what accuracy these parameters should be estimated in real-world applications. The effects of systematic changes in p and/or θ in the l_p-norm are systematically analysed on the k-median results, using regular as well as irregular lattices of points. Different exponents are also applied to distance. All simulations lead to the following conclusion: changes in p, θ and α influence the absolute values of the location parameters (such as the maximum or the average distance) but have little effect on the location-allocation results (the choice of the location sites).

Practical implications arising from these simulations are twofold. Firstly, small errors on the empirical estimation of distances by the l_p-distance predicting function very slightly affect the k-median results. This is particularly true for values of p close to 2 ($p \in [1.5\text{-}2.3]$). This result is of prime importance when considering the amount of work involved in estimating the real-world distances by l_p-distances: it saves data–collecting and computing time whenever such applications are considered. In other words, the substantial volume of extra work involved in determining the exact parameters for the l_p-distance estimation function is not really worth while in location-allocation modelling as it gives only a slender chance of obtaining more accurate or different results. Of course, this conclusion does not hold if the value of the p, α or θ parameter differs for each link of the network.

Secondly, Chapter 3 has already shown the significant influence of the shape of the network on the k-median results. Papini (1994) has also shown the influence of extreme points to locate centres. Given these two results and the results suggested in this chapter, one can say that the coordinates of the points and hence the shape of the transportation network have more importance for the k-median results than the way the distance is measured.

These conclusions are of prime importance in planning and real–world applications. They are of course limited by the conditions of the experiments. Further simulations should be performed in order to see if other optimal location-allocation models are evenly insensitive and if other spatial models involving a measure of distance lead to the same kind of conclusions (such as transport, gravitation, centrality, and so on).

7. Price policies, transportation networks and location-allocation results

7.1 INTRODUCTION

In this chapter, we are concerned with the financing of the transportation costs. The price of a good or a service produced in j and consumed in or by the clients in i depends upon the total production costs in j as well as upon the transportation costs between the commodity j and the client i. Transportation costs can be paid by the firm, by the clients, by other clients or by a combination of them. Greenhut (1981) observed that about three–quarters of the firms surveyed in the US, West Germany and Japan practice some form of delivered pricing, thus bearing some – or all – of the transportation costs. Spatial price policy is encountered in the real world and consists in determining one or several delivered prices together with the zones in which they apply. Hence, if the pricing policy is introduced in location-allocation models, demand is no longer exogenous to the model: it depends upon fixed production costs, upon the total number of locations and upon the financing of the transportation costs.

A major issue of economic geography is to predict the possible impact of public policies on regional development. Public authorities have at their disposition a broad spectrum of tools of different natures for influencing the decisions of economic agents. Legal constraints can limit the range of possible actions of these agents, while financial incentives in the form of taxes or subsidies are supposed to encourage or discourage some types of behaviour. Typical of the former is the bundle of laws issued in different countries regulating the geographical systems of prices that can be followed by private and public firms. Phlips (1983) and Greenhut, Norman and Hung (1987) provide insightful discussions on this topic. Another way of reaching regional or national objectives is through the provision of public services and infrastructures. One type of infrastructure which is likely to have a marked influence on the spatial organisation of the society is the communication network.

In Chapters 3 to 5 we have tried to evaluate the impact of the design of a transportation network on the location of economic activities. The main finding is that different families of shapes of networks induce very different location patterns, hence that the decisions regarding the general form of communication networks and the modifications brought to them are of strategic importance to regional development. Those simulations were first carried on using the *k*-median problem (Chapter 3). In Chapter 4, we performed similar analyses using the uncapacitated plant location problem in order to investigate the joint influence of the network morphology and of economies of scale. In the present chapter, we elaborate further on this topic by also introducing the spatial pricing policy followed by the firm, that is, the way total distribution cost is shared among the clients of the firm. Our purpose is to try to disentangle the relative influences of the shape of the transportation network, of economies of scale in production, and of the geographical pricing system on the location pattern of producing facilities of a firm, on its spatial schedule of prices, and thus on the characteristics of its market area. Indeed, most of the analysis of the geographical systems of prices assume a given fixed pattern of producing facilities and focus on the properties of the market areas. The question of whether different pricing systems generate different configurations of plants is thus in order.

Although the elaboration of integrated price–production–location models has been advocated long ago (Greenhut, 1956), it is only recently that they have been seriously tackled in the literature. In recent years, there have been computational and theoretical advances in the analysis both of the equilibrium and of the disequilibrium properties of pricing models (Plummer, 1996a and 1996b; Mulligan, 1996; Nagurney and Zhao, 1993; Nagurney, 1993). Examining the literature in spatial economics, one can roughly say that the geographical aspects of the organisation of production have been traditionally reserved to the conventional Weberian approach of location theory, while the pricing questions were considered in the Hotelling framework of spatial competition. A series of papers (Hanjoul et al., 1990; Hansen et al., 1995; Hansen et al., 1997), have proposed models and algorithms to solve the problem of the spatial monopolist. Peeters and Thisse (1995) as well as Peeters (1995) provide a synthesis of these approaches. The paper written by Peeters and Thisse in 1990 was a first attempt to appreciate the geographical impacts of some pricing policies. The present chapter develops this; it considers alternative price policies and several types of transportation networks.

This chapter is organised as follows: in Section 7.2, we introduce the general pricing–production–location problem of the spatial monopolist and we particularise it for the five spatial pricing policies considered in this analysis. In Section 7.3, the protocol of the computational experiments is described. In Section 7.4, the results of the empirical tests are presented. We conclude with a brief discussion of our results in Section 7.5.

7.2 METHODOLOGY

7.2.1 The Price Location Problem

The problem consists in finding the best location and allocation patterns of private firms, given a fixed production cost, in order to maximise profit. This problem is conditional on the price policy which defines the optimal size of the market area. It will be assumed that the firm operates as a spatial monopolist. This means that the firm is the sole producer of the commodity considered, or that its potential competitors are settled far away and/or transportation costs are high enough to prevent an invasion of the firm's market area, or that there exists an agreement between firms allowing each of them a given territory that will not be encroached on by the others. Consequently, the firm is free to choose at which price it sells its product. It is also assumed that the clients are uniformly spread with unit density. Comforted by the assertion of Greenhut, Norman and Hung (1987), each client is endowed with a linear demand function.

A spatial monopolist produces a single commodity within a region. Following the notations used by Peeters (1995), each client i $(i = 1,\ldots,I)$ has a linear demand function $D_i(\pi) = [a_i - b_i\,\pi]^+$, where π is the delivered price at the location of client i and $[x]^+ = \max(x, 0)$. Let $\alpha_i = a_i/b_i$ be the maximum willingness to pay of client i. The firm can operate up to J plants located at sites $j = 1,\ldots,J$. The fixed cost for opening a plant at site j is f_j, while the local constant marginal production cost is v_j. The unit transportation cost for supplying client i from site j is denoted by t_{ij}.

Three sets of variables were also introduced. Firstly, the variable π_{ij} is the delivered price charged by the firm for supplying client i from site j. Secondly, y_j is set equal to 1 if a plant is operating at site j and to 0 otherwise. Thirdly, the variable x_{ij} is set equal to 1 if client i is supplied from site j and to 0 otherwise.

The profit of the firm is equal to its revenue, minus the production costs, minus the distribution costs. Hence the objective function can be stated as follows:

$$P(\Pi, y, X) = \sum_{i=1}^{I}\sum_{j=1}^{J}\pi_{ij}D_i(\pi_{ij})x_{ij} - \sum_{j=1}^{J}(f_j y_j + v_j \sum_{i=1}^{I} D_i(\pi_{ij})x_{ij})$$

$$- \sum_{i=1}^{I}\sum_{j=1}^{J} t_{ij} D_i(\pi_{ij})x_{ij}$$

$$= \sum_{i=1}^{I}\sum_{j=1}^{J}(\pi_{ij} - v_j - t_{ij})D_i(\pi_{ij})x_{ij} - \sum_{j=1}^{J} f_j y_j$$

$$= \sum_{i=1}^{I}\sum_{j=1}^{J} B_{ij}(\pi_{ij})x_{ij} - \sum_{j=1}^{J} f_j y_j$$

where $B_{ij}(\pi_{ij})$ can be interpreted as the gross benefit of the firm for supplying client i from site j at the delivered price π_{ij}. The general price-location problem (PLP) of the spatial monopolist is to determine the price charged to each client, a configuration of production facilities, and an allocation of the clients to those facilities in order to maximise its profit (see Hanjoul et al., 1990; Peeters and Thisse, 1990). A possible formulation is the following:

$$P^* = \text{Max } P(\Pi, y, X) \tag{7.1}$$

subject to the following contraints:

$$\sum_{j=1}^{J} x_{ij} \le 1, \qquad i = 1,\dots,I \tag{7.2}$$

$$x_{ij} \le y_j, \qquad i = 1,\dots,I; \qquad j = 1,\dots,J \tag{7.3}$$

$$(\Pi, X) \in \Pi \times X \tag{7.4}$$

$$\pi_{ij} \ge 0 \quad \text{or} \quad 1, \quad i = 1,\dots,I; j = 1,\dots,J \tag{7.5}$$

$$x_{ij} = 0 \quad \text{or} \quad 1, \quad i = 1,\dots,I; j = 1,\dots,J \tag{7.6}$$

$$y_j \ge 0 \quad \text{or} \quad 1, \quad j = 1,\dots,J \tag{7.7}$$

The first group of constraints (7.2) implies that each client is allocated to at most one facility; it simultaneously ensures that no client receives more than its demand and that the firm is not obliged to supply a client when this entails a loss. The second group of constraints (7.3) prevents any allocation to a site where no plant has been opened. The third set of constraints (7.4) represents in a very general way the spatial pricing policies that can be followed by the firm; it specifies the geographical distribution of the delivered prices and the ensuing distribution pattern for the clients. Finally, the positivity constraints (7.5) on the variables representing the prices and the integrality constraints (7.6) and (7.7) on those representing the locations of plants and the allocations of clients to sites are produced.

More details about this general model, its particularisation for the five geographical pricing systems used in this chapter and discussed in the following section, and related algorithmic resolution issues, can be found in Peeters (1995) and in Peeters and Thisse (1995).

7.2.2 The Price Policies

The literature on spatial pricing is abundant. Our aim here is to define and to justify our choices before introducing them in the location problem. For each client, there is a mill (or f.o.b.) price (here noted *p)* which is the price of the good set by the firm at the place of production j, and a delivered price which is equal to the mill price plus the transportation cost between j and i ($\pi(r)$ is the delivered price at location $r \geq 0$). The particular shape of $\pi(r)$ as a function of r defines the spatial price policy of the firm, that is to say, the way in which transportation costs are passed on to the clients. Hence the optimal size of the market area depends on the spatial price policy (Peeters and Thisse, 1996). There exist many price policies; our choice is here limited to five price policies that are the most likely to be followed by a spatial monopolist. Let us give their definitions on grounds of one-dimensional geographic criteria (Peeters, 1995, pp. 134–46).

The discriminatory pricing policy

Let us first assume that the firm follows a non-uniform discriminatory pricing policy (denoted in the sequel by D). In this context, the firm determines for each pair of clients and sites the optimal price and the resulting potential gross benefit:

$$B_{ij}^{D} = B_{ij}(\pi_{ij}^{D}) = \mathrm{Max}(B_{ij}(\pi_{ij}), \pi_{ij} \geq 0), \quad i = 1, \ldots, I; \quad j = 1, \ldots, J$$

Hansen and Thisse (1977) observed that the general problem can be simplified to yield an uncapacitated facility location problem (UFLP).

With this policy, the price depends only upon the location of the client and not on the size of the market area. Moreover, insofar as the clients do not have access to a (much) cheaper transportation system, it will not be interesting for them to proceed to the transfer of commodities among places, and the firm has a natural way of separating the markets, a necessary condition to implement a discriminatory policy efficiently. If this price policy presents some advantages from a theoretical point of view (Holahan, 1975; Phlips, 1983), it also presents some notorious drawbacks: the list of prices is far from being transparent to the clients, a relocation of the firm entails a complete recalculation of the prices, bookkeeping may turn out to be very burdensome when the number of clients is high, it is hard to use it during promotional campaigns and it is illegal in several countries.

The uniform delivered pricing policy

In this case, the firm sets a mill price, which is identical for all clients, regardless of their location. This occurs when the clients travel to the plant and take care of the transportation, or when the firm delivers the commodity and charges its clients their respective transportation costs (Beckman, 1976). This price policy is generally considered as the only one that does not involve discrimination among clients and for that reason has been advocated by antitrust authorities.

Here, the firm may charge a uniform delivered price (U), independent of the location of the clients. This can be expressed by the constraints:

$$\pi_{ij} = \pi^U, \quad i = 1,...I; \quad j = 1,...,J$$

For any fixed value of π^U, this turns out to be a UFLP, so that the procedure is amenable to finding the global optimum of a univariate function defined by an oracle.

The uniform mill pricing policy

When the firm opts for a uniform mill pricing policy (*M*), the constraints are particularised as:

$$\pi_{ij} = p^M + t_{ij}, i = 1,...,I; j = 1,...,J$$

$$\sum_{\{k|t_{ik} \le t_{ij}\}} x_{ik} \ge y_j, i = 1,..,I; j = 1,...,J$$

It is generally assumed under this policy that each client travels to the nearest open facility, hence bearing the transportation cost he/she generates. The first group of constraints expresses this. Now, if the marginal production costs are not uniform across the sites, then the decisions of the clients are not necessarily optimal from the point of view of the firm. The second set of constraints states that the decisions of the clients override the wishes of the firm.

The mixed (pick-up or delivery) policy

In the mixed policy (m), the clients are free to choose between travel and pick-up at the plant at a mill price p^m or home delivery at a uniform delivered price π^m (Furlong and Slotsve, 1983). The mathematical expression of a client's choice is then more complex:

$$\pi_{ij} = p^m + t_{ij}, \textit{ if } B_{ij}(p^m + t_{ij}) > 0 \textit{ and either } B_{ij}(\pi^m) = 0 \textit{ or } B_{ij}(\pi^m) > 0 \textit{ and } p^m + t_{ij} \leq \pi^m$$

$$= \pi^m, \textit{ if } B_{ij}(\pi^m) > 0 \textit{ and either } B_{ij}(p^m + t_{ij}) = 0 \textit{ or } B_{ij}(p^m + t_{ij}) > 0 \textit{ and } p^m + t_{ij} \geq \pi^m$$

$$= \infty, \textit{ if } B_{ij}(p^m + t_{ij}) = B_{ij}(\pi^m) = 0$$

This expression states that if client *i* is supplied by a facility in *j*, this client will choose pick-up if this service is available (p^m is high enough to yield positive profit to the firm), if $p^m + t_{ij}$ does not exceed the maximum willingness to pay of the client, and if either home delivery is not provided at the client's location or it is possible but at a delivered price π^m that exceeds the full price $p^m + t_{ij}$ of self-service. The conditions for choosing home delivery are similar and we must also take into account the possibility that the client has neither access to one service nor to the other in which case we conventionally set the price to infinity. Furthermore, we also have to append constraints expressing the allocation rules: in case of pick-up, the client patronises the nearest open facility, while in case of home delivery the firm supplies from the facility accruing the largest benefit. This leads to the following formulation:

$$\sum_{\{k|k \text{f}_i j\}} x_{ik} \geq y_j, i = 1,\ldots,\text{I}; \text{j} = 1,\ldots\text{J}$$

where $k \ \phi \ j$ means that site *k* is potentially preferred to site *j* for supplying client *i*, that is, either $\pi_{ik} < \pi_{ij}$ or $\pi_{ik} = \pi_{ij} = \pi^m$ but $B_{ik}(\pi^m) > B_{ij}(\pi^m)$.

The zone pricing policy

In this case, the clients of the firm are clustered into different zones and the firm sets a uniform delivered price in each zone. The zones can be defined endogeneously or given a priori (for example in order to match administrative boundaries). In other words, the firm chooses intra–zonally the same delivered price, but inter–zonally different delivered prices as well as the boundaries of each zone (Peeters and Thisse, 1996). Let *Z* be the number of zones, fixed exogeneously; the client located at r is allocated to the zone z when the benefit that the firm can earn by placing him there exceeds any other allocation:

$$\pi(r) = \pi^{z} \quad \text{if} \quad B_r(\pi_x^Z) = \max(B_r(\pi_{z'}^Z) | z' \in Z)$$

π_z^Z being the price prevailing in zone z.

It can be shown that both the consumers as a whole (but not each of them) and the firm are better off when the number of zones increases. These positive effects are due to the fact that more and more consumers are supplied as the number of zones extends. When only one zone is considered, we fall back on the results of uniform delivered pricing while perfect spatial discrimination prevails when the number of zones becomes arbitrarily large. The zone pricing policy is a close substitute for perfect spatial discrimination. This system is viable only if the firm can hermetically seal the zones against transfers of commodity among clients, since otherwise the clients located at the outskirts of a zone could profitably resell to clients of the neighbouring zone at a lower price than the delivered price charged by the firm. Note that when only one zone is considered ($Z = 1$), this price policy is equivalent to the uniform delivered pricing U described before, and when the number of zones is very large ($Z \to \infty$), it is equivalent to the perfect spatial discriminatory pricing D. In our case, Z will be set to 2.

Comparison of the market areas

Before introducing price policies in the PLP model and testing the sensitivity of the optimal location-allocation to their variations, it is first worth comparing the expected market areas in a one-dimensional space. This one-dimensional comparison helps us to better understand the simulations discussed in Sections 7.3 and 7.4. Further developments regarding these one-dimensional aspects of the price policies can be found in Peeters (1995), pp. 143–6.

Figure 7.1 refers to the comparison of M, D and U in terms of market areas; it shows that on the first half of the [0; 2/3] interval, the clients are better off under M, while those located in the second half prefer U. The market area is 50 per cent larger under D. Even if the extension of the mill and uniform pricing policies are equivalent, different geographies of price and demands of commodity are observed. Compared to the mill pricing, the firm captures more surplus of the consumers located close to the plant while the remote clients are better off. U corresponds to the most widespread form of discrimination. The most important drawback of this policy is that there is no arbitrage between supplied and non-supplied areas.

The surplus of nearby customers is very high under *M*; under *U*, it is seriously pinched but remote clients do not benefit entirely from this redistribution (Peeters, 1995, p. 144). It explains why the total surplus is lower under *U* than *M*. Under *D*, the firm does not encroach so much on the surpluses in the vicinity of the plant and remote clients in the range [2/3; 1] benefit from a small surplus.

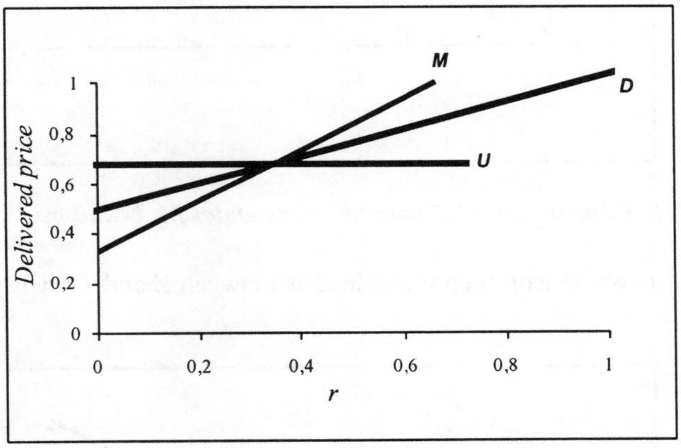

Figure 7.1 Comparison of delivered prices under M, U and D

The comparison between *M*, *U* and *m* pricing systems is reported in Figure 7.2. The mixed pricing policy (*m*) allows an extension of the market area. This is obtained by an increase of both the mill price and the delivered price: the firm separates nearby and remote clients and gains on both sides. In terms of consumers' surplus, it can be shown (Peeters, 1995, p. 145) that nearby clients are worse - off under *m* than under *M*, whereas remote clients lose compared to *U*, but more clients can be supplied and thus benefit from a surplus.

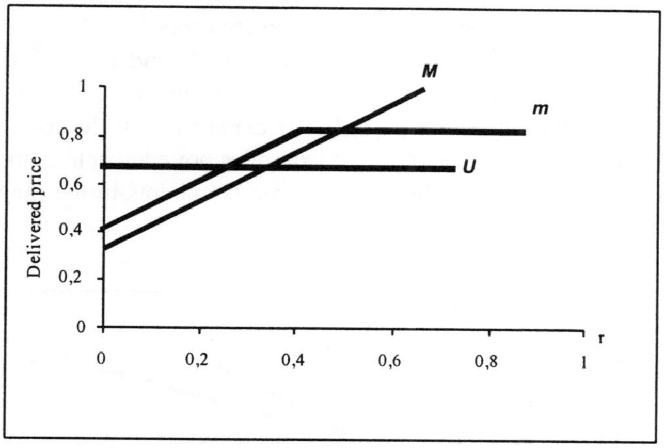

Figure 7.2 Comparison of delivered prices under M, U and m

The last comparison mentioned here is between *Z* and *D*; it is reported in Figure 7.3.

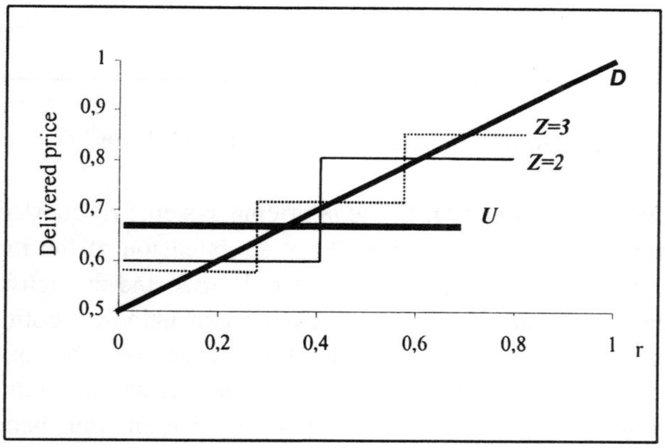

Figure 7.3 Comparison of the delivered prices under U, D, Z = 2 and Z = 3

It allows us to see the evolution of the one-dimensional geography of prices when passing from a unique zone (*U*) to *Z* = 2, *Z*= 3 and to *D*. The most striking feature is how the multiplication of the zones permits the conquest of more remote markets. Peeters (1995, pp. 207-27) demonstrated that zone pricing has many advantages, since it permits the conquest of a larger market area than under uniform delivered pricing and allows the firm to define a small number of tariff zones to obtain performances that are similar to those of spatial discriminatory pricing. Peeters and Thisse (1996) have also shown that zone pricing is a good approximation of discriminatory pricing and therefore may be chosen by firms because of its greater simplicity.

Our aim is now not to find the optimal pricing policy but to see how far the location-allocation results depend upon the price policies and how they vary with transportation networks. As discussed by the one-dimensional descriptions and in the figures proposed in this section, we expect spatial differences in a two–dimensional space. Let us apply the model described in Section 7.2.1 for the five price policies presented in Section 7.2.2. Section 7.2.3 defines the experimental design.

7.3 PRINCIPLES OF THE COMPUTATIONAL EXPERIMENT

Computational experiments are conducted in order to evaluate the relative influence (1) of the spatial pricing policy, (2) of the magnitude of the fixed costs, and (3) of the shape of the communication network 'channelising' the movements of goods and persons. Regarding the last of these, we have once again decided to operate on idealised networks rather than real-world layouts, in order to isolate the influence of a particular morphological characteristic from many other possible ones. Furthermore, working with a hypothetical settlement enables us to conduct tests without the additional complexity introduced by empirical modelling, as it is completely free from any contamination related to data collection, coding or others errors, and should lead to a greater generality of the conclusions.

We have performed simulations on a discrete transportation space, with a multiplant monopolistic firm (see preceding section). The clients of the firm are supposed to be located at the vertices of a 9×9 squared lattice (note that the size of the network is highly constrained by computational limitations). Each vertex is also a potential supply site. It is characterised by its coordinates (x_i, y_i), which are integer multiples of 100.

The theoretical transportation networks are constructed by joining some of those vertices. Each network is characterised by the number of edges and by their spatial organisation. The length of an edge is equal to the Euclidean distance between its endpoints. Four transportation networks are designed and used in the tests. Their choice is highly oriented by preceding results: two regular (*Eucl*, *Rect*) and two radial networks (*Rad*, *Circ*). Network *Eucl* corresponds to a complete graph (not illustrated). In network *Rect*, only the horizontal and vertical edges are considered (no oblique link). Radial networks have fewer links than the rectangular network. *Rad* represents a radial network without any peripheral link. It is well known that this type of network tends to favour the centre for accommodating an optimal location (Perreur and Thisse, 1974). When adding a peripheral link to a radial network, one obtains a circumradial network (*Circ*). In this type of network, accessibility is theoretically increased (more edges); Perreur and Thisse showed that the optimal locations of the supply sites may also occur on the peripheral road, more particularly at the intersection with the radial axes. In Chapter 3, we also showed that the position of the peripheral road is of prime importance to location results; the location of the ring road is here median: it is governed by those preceding results.

On each network, we apply the five variants of the PLP described in Section 7.2. We assume a common fixed cost F across the locations. In Chapter 4, we have attempted to show the joint influence of F (which may represent the economies of scale in production costs), and of the shape of the network on the solution of the uncapacitated facility location problem defined on a similar setting. F is here set equal to 2 700 000; in Section 7.4.3 results are discussed for a smaller value of F. We also assume that the marginal production cost is set to zero for every potential site.

Finally, we assume that the demand function of the clients is affine with an intercept of 2000 and a slope equal to 1: $D_i(\pi) = [2000-\pi]^+$, $i = 1,\ldots,I$.

7.4 EXPERIMENTAL RESULTS

Let us first comment in Section 7.4.1 on the experimental results with a relatively high fixed cost (F = 2 700 000), a circumradial network, and the five pricing policies (see preceding sections for choices and definitions). We then consider other networks (Section 7.4.2), and later, comment on alternative experimental conditions (Section 7.4.3).

7.4.1 Price Policies and Circumradial Networks

Let us comment on the results with a high fixed cost (F = 2 700 000), and a radial configuration with a peripheral ring road. A high fixed cost means high economies of scale, and hence few optimal locations (k small). In our case, the model suggests a k value equal to 4, whatever the price policy. Hence, in our case, the price policy does not affect the number of optimally located commodities.

Table 7.1 gives a general comparison of the results according to the five price policies. Profit values show very slight differences; the highest value is observed for the D policy, intermediate values for policies m and Z, and the smallest values for U and M. This confirms the equivalence of profits established by Beckmann and Ingene (1976). Average transportation costs are also very similar with the highest value observed for *U* and the smallest for *M*. This is in accord with Beckmann (1985). Plant capacities are similar, whatever the price policy. Optimal locations are summarised in Figure 7.4 and Table 7.1. Optimal locations are always chosen at the crossing between the radial axes and the peripheral ring road, whatever the price policy. The price policy does not generate differences in the optimal locations, at least in this toy-network.

A service area is represented by a spatial set of demand points allocated to a selected supply site. Differences in allocation are observed in the case of the mill price policy (*M*) for points located at the border of the market (service) areas; all other pricing policies lead to exactly the same market areas. Hence we can conclude that in our quite simple experimental framework the price policy only slightly affects the shape of the service area. Note that this result depends strongly on the assumption of equal marginal production costs among potential sites; when they differ markedly, price policies *M* and *m* yield areas which are distinct from those resulting from other policies (see for example Peeters and Thisse, 1995).

In our simplistic studied configuration, the number of facilities and their location do not vary with price policy, and the shape of the market areas is relatively stable.

Table 7.1 PLP results with a Circ network

	D	*U*	*M*	*m*	*Z*
Capacity of					
site 1	14 569	14 366	14 772	14 484	14 484
site 2	17 967	17 967	17 977	17 964	17 963
site 3	17 967	17 967	17 977	17 964	17 963
site 4	22 223	22 446	22 001	22 313	22 312
Profit (x 1000)	54 677	54 498	54 498	54 618	54 618
Average transportation cost	198.4	203.3	193.5	199.3	200.0

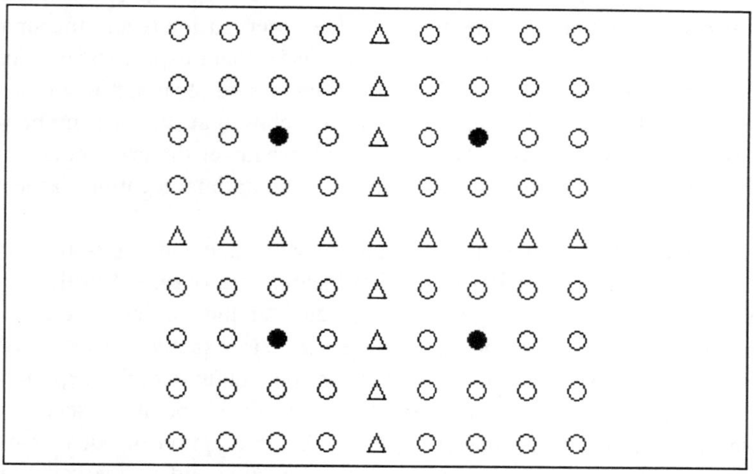

Note: dark dots: optimal locations; triangles: demand points allocated differently with the price policy; white circles: demand points with stable allocation

Figure 7.4 Optimal locations on a radial network, whatever the price policy

Now, what about the quantity of demand generated in each demand point? We know that this quantity is endogenous, depending upon the price policy (Section 7.2). Table 7.2 displays some simple descriptive statistics about the observed differences. The total quantity of demand does not vary with the price policy; in the circumradial case (*Circ*), it is always equal to 72 727, but the distribution of the individual values of demand depends upon the price policy. As expected, the uniform price policy *U* generates the same quantity of demand in each selected supply site; the policies *D*, *m* and *Z* exhibit an average spatial variation; while the policy *M* has the largest variation in space: under the mill policy, demand is highly discriminated (for the rationale see for example Peeters and Thisse, 1995).

The demand values observed at the 81 points for the five price policies are highly and significantly correlated (Table 7.3): as illustrated on Figure 7.5, their variation in space is rather similar, but the mill pricing (*M*) shows the most important spatial variations. Pearson correlation coefficients are computed between the spatial distributions; all coefficients are significantly ($\alpha = 0.0001$) different from zero and are higher than 0.8, whatever the price policy; increases and/or decreases of demand occur at the same places (Table 7.3). Figure 7.6 illustrates where the differences occur on a map; given the preceding results; it is restricted to the comparison of mill and uniform pricing. Mill pricing creates a clear-cut hierarchy of places: the locations of the commodities concentrate much more demand in their close vicinity than the less accessible points. The map shows that the observed differences reinforce the radial axes: spatial structure and price policy interact. Explanations are to be found in Peeters (1995, p. 201): the firm bears the total transport costs in the *D* and *U* cases, while transportation costs are entirely passed on to the customers in the *M* case. In the first two cases, the firm internalises the whole (social) benefit generated by the choice of an efficient configuration.

Table 7.2 Descriptive statistics about the quantity of demand generated at each of the 81 demand note

Network & pricing policy	Total demand generated	Standard deviation	Minimum	Maximum
Circ				
U	72 727	0.00	898	898
D	72 727	47.24	800	1000
M	72 727	94.46	701	1101
m	72 727	46.52	870	1048
Z	72 727	38.77	870	951
Rad				
U	66 499	0.00	821	821
D	66 500	66.74	718	1000
M	66 500	133.47	614	1178
m	66 500	72.87	779	1103
Z	66 499	56.15	779	896
Rect				
U	72 160	0.00	891	891
D	72 160	50.49	800	1000
M	72 159	100.98	708	1108
m	72 162	65.59	838	1076
Z	72 160	41.71	838	923
Eucl				
U	73 956	0.00	913	913
D	73 957	34.33	858	1000
M	73 957	68.66	804	1086
m	73 957	40.99	887	1053
Z	73 955	29.62	887	946

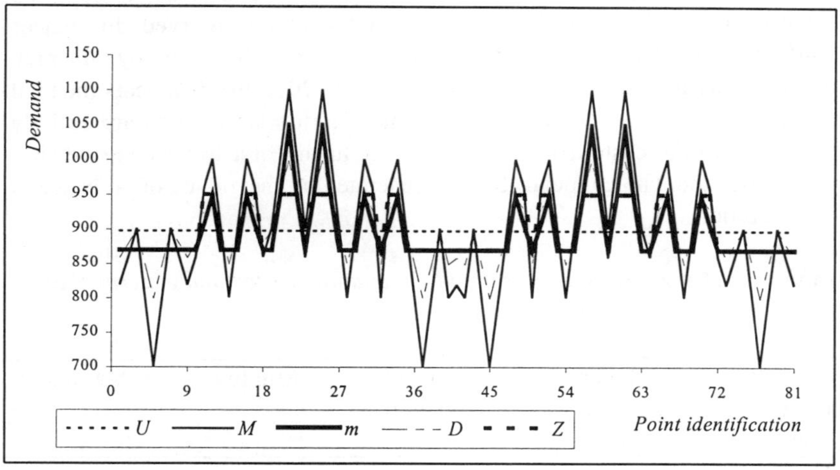

Figure 7.5 Variation of the quantity of demand generated in each demand point i (1,...,81) by each pricing policy (U, M, m, D and Z)

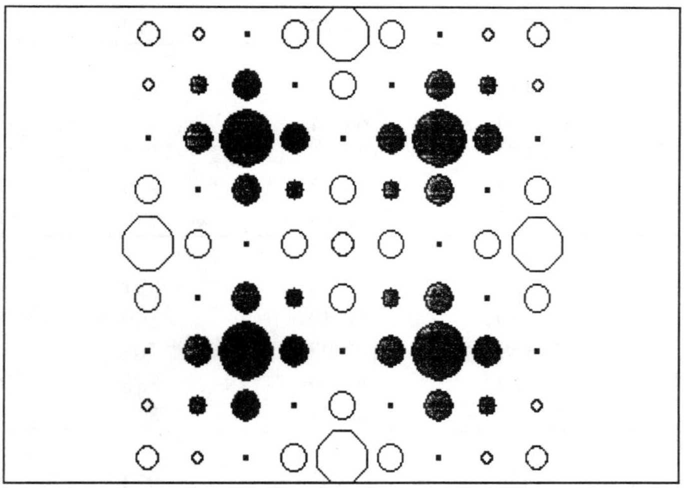

Note: circle proportional to the observed difference; grey: increase, white: decrease

Figure 7.6 Difference between the quantity generated by the mill (M) and uniform pricing (U) policies in each of the 81 points of the lattice for the Circ network

We can here conclude that – all other things being equal and within the limits of the studied toy-networks – the PLP results are only slightly sensitive to changes in price policy: the same number of selected facilities, the same locations, slight differences in the shape of the market areas only for mill pricing and the same global quantity of generated demand. Differences are only observed in the quantity of demand generated in each demand point, thus in the geographical distribution of demand.

In order to check our results, we will comment on some further simulations on other spatial networks (Section 7.4.2), or with another value of F (Section 7.4.3).

Table 7.3 Pearson correlation coefficients between the amount of demand generated in each i for each pricing policy (network: Circ)

	M	*m*	*D*
m	0.833		
	0.0001		
D	1.000	0.833	
	0.0001	0.0001	
Z	0.819	0.832	0.719
	0.0001	0.0001	0.0001

Note: italic = significance level

7.4.2 Further Simulations on other Networks

Let us now extend the preceding simulations to other spatial configurations designed on the same squared lattice of points, more precisely to regular networks (*Rect*, *Eucl*) and to the radial network with no peripheral (*Rad*) described in Section 7.3.

Table 7.4 gives the number of optimally located facilities for each price policy and each spatial configuration. It confirms that the number of optimal locations is not affected by the price policy. The locations of the commodities show some slight differences with price policy for regular networks: the observed differences are mainly due to local optima resulting from symmetries in the networks. The spatial homogeneity of the regular network favours local optima; a simple rotation of the spatial results shows that the solutions are identical. Allocations are also slightly affected by changes in price policy: the observed differences are bound to the local optima. The total quantity of demand generated is not affected by the price policy whatever the underlying spatial structure (Table 7.2) and the individual spatial variation is the same as the one previously described (see Section 7.4.1): the highest variation is observed for the mill pricing.

It is worth noting here that the effects of the price policy yield for a given spatial structure, and that the shape of the network generates important differences. They are summarised in Tables 7.2, 7.4, 7.5 and in Figure 7.7. Table 7.2 shows that the total amount of generated demand varies with the network shape: *Eucl* generates the largest amount of demand because of the highest underlying accessibility. *Rect* and *Circ* have similar values of demand confirming preceding results (Chapter 3): when a peripheral ring road is optimally located, the spatial environment functions as well as denser regular networks. The radial network *Rad* generates a significantly much lower quantity of demand: in our quite simple configurations, the efficiency of the underlying transportation network plays a much more significant role than the price policy. Table 7.4 shows that the number of optimally located facilities is always the same except for the radial network: the radial network privileges the central location in such a way that only one facility is located. Table 7.5 gives the correlation coefficients between the spatial distributions of demand generated by one price policy (*M*) according to the different networks. Correlation coefficients are not as high as those observed between price policies for a fixed network (Table 7.3) and they are often not significant.

Table 7.4 Location results for 5 price policies and 4 spatial structures when F = 2 700 000

	D	*M*	*m*	*U*	*Z*
Rect	**4**(1) 20-25-57-70(2)	**4** 16-21-56-61	**4** 20-25-57-70	**4** 16-21-56-61	**4** 16-21-56-61
Eucl	**4** 20-25-57-70	**4** 20-25-57-62	**4** 20-25-57-62	**4** 20-25-57-62	**4** 12-65-57-62
Rad	**1** 41	**1** 41	**1** 41	**1** 41	**1** 41
Circ	**4** 21-25-57-61	**4** 21-25-57-61	**4** 21-25-57-61	**4** 21-25-57-61	**4** 21-25-57-61

Note : (1) number of optimally located facilities
(2) index of the optimal location

Table 7.5 Pearson correlation coefficients between the spatial distributions of demand generated by the Mill policy on different networks

	Eucl	*Rect*	*Quad*
Rect	0.559 [(a)]	-	
	0.000[(b)]		
Rad	0.093	0.105	-
	0.409	0.352	
Circ	0.747	0.756	0.235
	0.000	0.000	0.034

Notes: [(a)]: correlation
[(b)]: significance level

Figure 7.7 gives a last illustration of the importance of the transportation network in demand distribution: for each network configuration, it shows the locations of the differences between the quantity of demand generated by the *M* and *U* pricing policies.

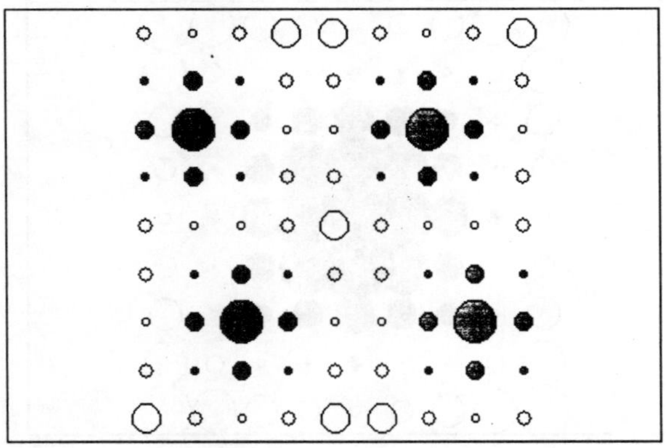

Notes: circle proportional to the difference
grey: increase, white: decrease.

Figure 7.7 Difference between the quantity generated by the M and U policies in each of the 81 points for the Euclidean network

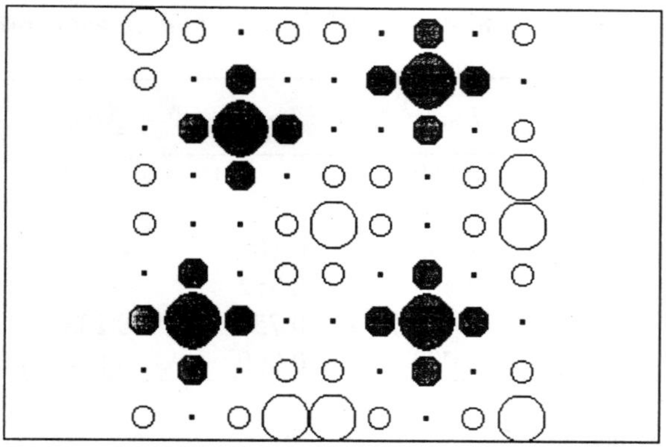

Notes: circle proportional to the difference
grey: increase, white: decrease.

Figure 7.8 Difference between the quantity generated by the M and U policies in each of the 81 points for the grid network

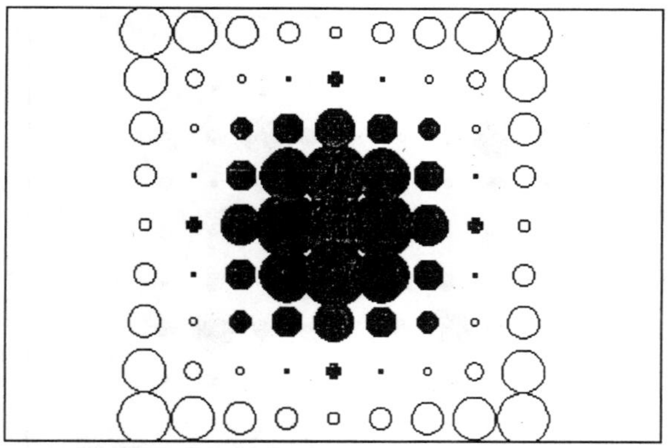

Notes: circle proportional to the difference
grey: increase, white: decrease.

Figure 7.9 Difference between the quantity generated by the M and U policies in each of the 81 points for the radial network

The extension of the simulations proposed on a radial network with a ring road (Section 7.4.1) to other networks (Section 7.4.2) has not changed our conclusions on the effect of price policy on location-allocation results but have added some insight about the important structural effect of the underlying transportation network.

7.4.3 Alleviating Some of the Simulation Constraints

It is never easy to draw conclusions from an analysis based on simulations, especially when several sources of variations are involved. Being aware of this, we have studied several variants of our data set. Firstly, experiments were carried out using a value of 1 000 000 for the fixed costs, hence allowing for more facilities to be opened. The results are summarised in Table 7.6. They confirm our preceding finding: the impact of the morphology of the network on the optimal locations is much stronger than the impact of the type of pricing policy.

Table 7.6 Location results for 5 price policies and 4 spatial structures when F = 1 000 000

	D	*M*	*m*	*U*	*Z*
Rect	**7**[1]	**7**	**7**	**7**	**7**
	5-20-26-41-56-62-77[2]	5-20-26-41-56-62-77	12-16-37-41-45-66-70	5-20-26-41-56-62-77	12-16-37-41-45-66-70
Eucl	**5**	**6**	**5**	**6**	**5**
	12-25-56-59-62	12-17-38-41-62-66	16-21-56-59-62	12-26-38-41-66-71	12-25-56-59-62
Rad	**5**	**5**	**5**	**5**	**5**
	21-25-41-57-61	21-25-41-57-61	21-25-41-57-61	21-25-41-57-61	21-25-41-57-61
Circ	**4**	**4**	**4**	**4**	**4**
	21-25-57-61	21-25-57-61	21-25-57-61	21-25-57-61	21-25-57-61

Notes: (1) refers to the number of facilities optimally located
(2) refers to the index of the points selected as location

Secondly, parallel experiments were conducted on hexagonal lattices of points on which several transportation networks were designed (Algoed, 1996). Although limited by the size of the studied lattice (91 points), each transportation network confirmed the results obtained on the squared lattice: both price policy and transportation network influence the location-allocation results but the transportation network generates more important differences than the price policy. Although more investigation is called for – especially with larger data sets – these preliminary results converge.

7.5 CONCLUDING COMMENTS

This chapter has considered the UFLP with price-sensitive demands when the firm follows different pricing policies. Several transportation structures have been designed on squared and hexagonal lattices of points. As in the preceding chapters, it is not easy to draw conclusions from the analysis of the simulation results, especially when the variation of several parameters is involved. The main difficulty is the robustness of the results obtained. However, being aware of this, we have studied several extensions of our basic simulations. Most results point in the same direction.

The principles that seem to emerge from our analyses can be summarised as follows: the selection of a price policy has an impact on the distribution of the commodity and on the related geography of prices, thus confirming Greenhut's thesis (1956). But, the choice of a price policy has a limited impact on firms' location compared to the transport network configuration. This conclusion reflects the findings of Chapters 3 and 4. Although one has to be careful in extending conclusions from rather simplistic toy-networks to real-world situations, this suggests that shaping the transport network proves to be a rather powerful instrument in spatial planning, especially in countries with limited infrastructures like the less developed countries (see Plassard, 1977 for some empirical evidence and Chapters 3 and 4 for extensive experiments assessing the impact of the transportation network on location choices).

Empirical evidence of our results can partially be found in Plummer, Haining and Sheppard (1998): although in a different modelling context, they showed that the price awareness of consumers and the accessibility of gasoline stations appear to be of significant influence in deciding where to purchase gasoline. The authors conclude that observed spatial price variations depend on both the service characteristics of the individual gasoline retailers and the accessibility or location advantage that each station enjoys within the spatial configuration of the urban market. The gasoline stations which are more accessible have larger choice sets and charge lower prices; they tend to attract the most sales from other retailers (the authors used a household survey data collected from St Cloud, Minnesota). Price and transportation conditions seem to be crucial in spatial economic organisation. The experiments conducted in this chapter show that transportation structure is far more decisive than price policy.

8. Negative externalities and location-allocation results

8.1 INTRODUCTION

Location-allocation models are recognised as valuable tools for providing insights to planners and decision-makers regarding the impacts and relative attribute of potential planning activities. Standard models optimally locate *k* facilities and allocate *m* demand points to them. These models often optimise a social welfare criterion such as the total transportation costs minimisation; in this latter case, distance is usually a physical distance measured on a transportation network. Many facilities, however, generate effects on the neighbouring physical and/or human environment. The economic concept representing these effects is often called 'externality' (Section 8.2). The introduction of externalities in location-allocation models implies changes in the formulations of the models as well as in the resolution methods: in addition to transportation costs, the model includes a minimisation–maximisation function of the externalities on the nearby environment. Externalities are often measured on a network, but diffuse all over the space. The values associated with the edges of the transportation network are no longer restricted to one physical distance, but they are assumed to correspond to a sum of some transportation costs and some pollution costs. The optimal solution is a compromise between transportation costs and pollution costs. These latter are measured on selected edges and nodes, but they spread on the entire studied area, following a totally different diffusion process than transportation costs. Both costs are unwanted by the users of the facilities: users want a facility as close as possible and they also want the negative effects they have to undergo as small as possible. The issues of accessibility to facility services and minimising pollution exposure to communities provide an interesting planning situation in that competing siting objectives are involved. In this sense, regional planning is necessary in order to identify equitable and reasonable long-term alternatives for polluting facilities. Thus, this chapter tries to show the role of transportation costs – and hence, the shape of the transportation network – when negative externalities are also incorporated in the location-allocation model.

Undesirable facilities have been termed either noxious or obnoxious (Murray et al., 1998). Noxious facilities are those that have a potential risk to public health and safety such as a nuclear power plant or a hazardous waste storage site. Obnoxious facilities have less of a health risk but are not necessarily more desirable than noxious facilities. Obnoxious facilities are defined as facilities useful to the population but generating negative externalities in the surrounding environment (landfills; activities with excessive noise levels resulting from equipment operation such as fire stations, schools, and so on; certain kinds of social service centres such as drug treatment facilities or prisons). Generally, the effects of spatial externalities decrease with the increasing distance from the source (Papageorgiou, 1978).

The location of obnoxious facilities has raised many fundamental problems in the last decades. Many authors have been concerned with the 'maximin' problem that locates one facility on the plane and maximises the minimum distance to some protected areas such as towns, agricultural lands or nature reserves. (Dasarathy and White, 1980; Drezner and Wesolowsky, 1983, 1985, 1996; Melachrinoudis and Cullinane, 1985, 1986a and 1986b; Karkazis and Karagiorgis, 1987; Karkazis, 1988). Hansen, Peeters and Thisse also produced a model in 1981 (1981a): they considered a feasible location set formed by the convex union of polygons; the location theorem is proposed for both the minimisation of the total nuisance cost and the maximal nuisance cost. A maxisum problem has been considered by Church and Garfinkel (1978), and Berman, Larson and Chiu (1985); in this case, the obnoxious facility is located on a network as far as possible from a given set of population centres. At a few exceptions (Melachrinoudis and Cullinane, 1986), most modelling approaches assume that the measure of effectiveness is a linear function of the distance from the new facility to the existing facilities.

Erkut and Neuman presented in 1989 a fairly comprehensive review of the existing literature about undesirable facilities. They limited themselves to location problems with maximisation objectives. Both single and multiple facility models are considered. Most of the cited models are quite abstract, involving a single objective and simple functions of distance. Erkut and Neuman mention that little research has been conducted on multi-objective models applied to this particular location problem and suggested in 1992 one way of solving the problem. Their new model is a three-objective mixed integer programming formulation where the objectives are to minimise total cost, minimise total opposition and maximise equity (Erkut and Neuman, 1992); their approach is based on minimising the maximum disutility of individuals, measured using a decreasing distance and increasing facility capacity function.

Modelling pollution spread has also engaged the attention of researchers because pollution does not diffuse just along road networks: air, water and soil diffusions are concerned and interfere with each other. Authors have tried to fully represent the impact of pollution on the environment and have elaborated some complex but realistic models (Karkazis and Boffey, 1994; Matzoros and Van Vliet, 1992; Papadimitriou and Karkazis, 1991).

The purpose of this chapter is not to represent the pollution impact in a realistic way, nor to construct an empirical model about pollution. This chapter aims at systematically testing the sensitivity of a commonly used location-allocation model to the introduction of an externality function in its objective function. In other words: how far are the location-allocation results sensitive to the addition of a negative externality criterion to the model's objective function? Are the optimal locations sensitive to pollution, and by how much? Are the changes continuous, or do they proceed by steps? How much do transportation costs matter compared to pollution costs? In our case, negative externalities are added to the transportation costs, and their effects on optimal location results are analysed for a given transportation network.

The remainder of this chapter is mainly drawn from Laurent, Peeters and Thomas (1997) and is divided into three parts. In Section 8.2, the design of the experiment is described, as well as the inputs of the simulation program and the model itself. In Section 8.3, the results of the simulations are given in terms of efficiency (Section 8.3.1) and optimal locations (Section 8.3.2). Finally, Section 8.4 is devoted to conclusions and possible extensions.

8.2 DESIGN OF THE EXPERIMENT

In order to assure unambiguous interpretation of what follows, this section contains some definitions of concepts / inputs applied in this chapter.

8.2.1 The Settlement System

One set of points is chosen among all possible spatial configurations: an 18×18 squared lattice in which each point i has the same spatial environment as every other point, whith the exception of the border points and the central point. The choice of the shape and size of the lattice is governed by the preceding results: it appears to be the most satisfactory compromise between feasibility, reasonable computing time and practical results. Hence, 324 points are taken into account, scattered regularly in the space. The middle point of the lattice is added to the set of 324 points in order to make a central location possible (n = 325). Given some other simulations performed on other lattices, this choice has very little impact on the final conclusions of our experiments. No changes in transportation structure are considered: a complete network is here chosen (horizontal and vertical links + diagonals).

As mentioned in the preceding chapters, each point i of the lattice is simultaneously a demand point and a potential supply site; each demand point i is characterised by its coordinates (x_i, y_i) and is linked by an edge to its closest neighbours. A link has a length of 100 (edges) or 141 (diagonals): all demand/supply points are equidistant, at the exception of the middle point (Euclidean space). Hence, it is assumed that these points i represent population or customers centres that rely upon the services of the undesirable facilities.

Transportation costs between demand and supply are proportional to the physical distance measured on the transportation network; pollution costs diffuse continuously and isotropically in the plane but are measured on discrete locations (nodes on the transportation network). Transportation costs and nuisances are here artificially measured in the same units and are additive.

There is no variation in the amount of demand associated with each population centre i ($a_i = 1, \forall i$). Demand is totally satisfied by the optimally located facilities and is inelastic to distance. Hence, the chosen settlement system is a homogeneous environment, with no variation in weights and no privileged route directions. We assumed that the population that would oppose the facility is also the population that will derive the benefits from it, either directly or indirectly. All demand points are potential supply sites. We assume that more than one potential facility location can be identified and that all of the locations and the affected demand points (population centres) are under a common jurisdiction. We also assume the size of the facility as given and equal for all locations.

8.2.2 The Externality Functions

Although the concept of external effects is widely used in economics, there seems to be no consensus on its exact definition and interpretation. However, it is commonly recognised that externalities are an important source of market failure. Their existence leads to a deviation from the neo-classical world in which the price mechanism takes care of socially optimal resource allocation (Pareto efficiency) (see Verhoef, 1994b, for a short review). The term 'externality' can globally be defined as the indirect effect of a consumption (or production) activity on the consumption (or production) possibilities available to some other consumer (or producer). The term 'indirect', means that the effect concerns an agent other than the one exerting this economic activity, and that this effect does not work through the price systems. This effect has first been analysed by Marshall (1890) (in Catin, 1985 and 1994) and by Pigou (1920). It can be defined as a side effect which is not reflected in the costs and prices of a particular economic activity. For example, a power station without proper pollution controls is reducing its own costs at the expense of the health and welfare of others (Healey and Ilbery, 1990, p 337). This represents a deliberate transfer of costs of production from the producer to society at large; the cost of the pollution is not borne by the polluter. The essential feature of an external effect is that the effect produced is not a deliberate creation but an unintended or incidental by-product of some otherwise legitimate activity (Mishan, 1971, p. 2). Externalities within the urban context have led to numerous theoretical contributions (Papageorgiou, 1978; Kanemoto, 1987, or Fujita, 1989).

Smith (1977, p. 90) emphasised that the concept of externality is intrinsically spatial. An obnoxious facility generates a nuisance (a negative externality) diffusing in the surrounding environment. This spread can be confined in a close boundary or can have an international significance (for example acid rain). Various types of impacts produce different proximity-sensitive zones: for example, noise effects from a highway tend to dissipate more quickly with distance than air pollution impacts. Most externalities (noise, smell, congestion, and so on) do in fact have a fairly localised impact and their effects are weakened by distance from the source (see for instance Buchanan and Stubblebine, 1962; Mishan, 1971; Chapman and Walker, 1987 or Verhoef, 1994a). Harvey (1971), Smith (1977) and Bale (1978) visualise this in terms of a spatial externality field, with a downward sloping externality gradient in all directions from the source. This externality field can be depicted in terms of a cone, where an even population distribution surrounds the source or a set of contours. Such uniform distance-decay functions can of course be modified to take account of, for example, a dominant directional effect from the source (such as downwind of a power station) or a secondary node of concentration (such as a chemicals factory) (Healey and Ilbery, 1990).

In many instances, nuisance decreases non-linearly when distance increases: Melachrinoudis and Cullinane (1986) assumed, for example, that nuisance is inversely proportional to the squared distance, whereas Erkut and Neuman (1992) experimented with a general nuisance function where distance was raised to some negative power. Other authors considered complex and realistic representation of pollution diffusion function (for example Boffey and Karkazis, 1993; Karkazis and Boffey, 1994; Matzoros and Van Vliet, 1992; Papadimitriou and Karkazis, 1991).

In this chapter, nuisance is supposed to decrease linearly with the distance from the facility whatever the spatial direction; we consider that the atmosphere does not interfere. This type of diffusion function is not bound to one type of polluting activity; it is chosen for its extreme simplicity. Most dispersion functions are often sublinear, but the exaggeration inducted by the linearity enables to isolate and to emphasise the influence of the externality on the optimal locations. This simplistic modelling situation can be viewed as quite unrealistic; it enables us, however, to generalise our results without obscuring the underlying structure, and without leading to intractable empirical problems. We believe that this hypothesis is relevant in order to investigate the sensitivity of the location-allocation to externalities.

In our sensitivity analyses, the externality decreases linearly with distance from the facility *j*, in any direction. The line that describes the decreasing spatial effect of the externality from its source to its outer limit is defined by two quantities: the intensity (*Int*) and the extension (*Ext*) (Figure 8.1). Both quantities determine the importance of the obnoxious effect or nuisance. The intensity gives the measure of the externality at its source *j* (for example chemical pollutant concentration, sound level, and so on). *Int* decreases with distance from *j*. The extension is defined as the distance from *j* beyond which the externality is equal to zero; *Ext* measures the spread of the nuisance.

Given the size of the studied lattice of points, we let *Int* and *Ext* vary from 0 to 800, with a step of 100. Only some of the many computed simulations will be presented in the following paragraphs.

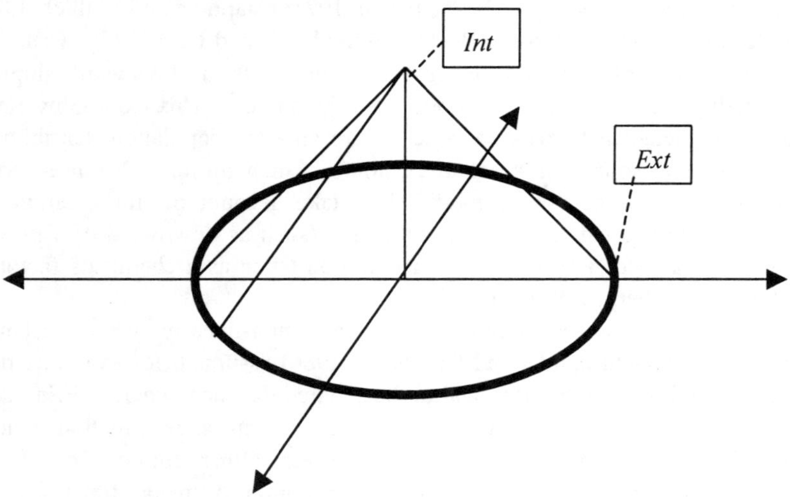

Figure 8.1 Intensity (Int), extension (Ext) and the shape of the studied externality: the externality cone and field

8.2.3 The Model

Although the location of undesirable facilities is clearly a multi-objective problem, there are only a few papers in the literature that treat it as such (see Erkut and Neuman, 1989 and 1992, for a review). Our aim is not here to incorporate the numerous trade-offs and conflicting concerns which arise when locating one or several undesirable facilities. The objective is to show the sensitivity of the optimal location to increasing levels of nuisance (compared to transportation costs); hence, the model is limited to a single objective and a simple function of distance.

A standard location-allocation model often minimises some formulation of transportation costs, optimally locates k facilities and allocates m demand points to them. The exact formulation of the objective criteria and of the constraints determines the type of modelling problem (for example Hansen et al., 1987; Labbé, Peeters and Thisse, 1995). Early models do not take into account the fact that facilities generate desirable as well as undesirable effects to the population within the spatial environment, and that the groups experiencing benefits are often different from the groups experiencing 'disbenefits'. This mixture of benefits and disbenefits characterises obnoxious facilities (Fonseca and Captivo, 1993).

We here aim at testing the effects of the introduction of negative externalities in a location-allocation model on the operational results, externalities being fixed costs in each point and to be added to the transportation costs. This addition may be seen as rather simplistic: other authors have solved this modelling problem by multicriteria models minimising transportation costs and externalities (for example Ratick and White, 1988; Erkut and Neuman, 1992; Murray et al., 1998 and others mentioned in Section 8.1). It has already been shown (Hansen, Peeters and Thisse, 1981a) that our approach is a special case of the bicriteria approach where the conversion factor between transportation cost and pollution objectives is assumed to be known.

A simple location-allocation model is here used (the k-median) and modified in order to include the externalities: the model minimises the sum of the transportation costs and the externalities characterising the impact of a facility.

This problem is an extension of the simple plant location problem (SPLP) containing a cardinality constraint; it can be written as follows:

$$\min F(x,y) = \sum_{i=1}^{m} \sum_{j=1}^{n} c_{ij} x_{ij} + \sum_{i=1}^{m} \sum_{j=1}^{n} E_{ij} y_j \tag{8.1}$$

subject to:

$$\sum_{j=1}^{n} x_{ij} = 1 \qquad (8.2)$$

$$0 \le x_{ij} \le y_j \qquad (8.3)$$

$$\sum_{j=1}^{n} y_j = k \qquad (8.4)$$

$$y_j \in \{0,1\} \qquad (8.5)$$

where

i demand point, i =1,...,m
j supply site, j = 1,...,n
m number of potential supply sites;
n number of demand points;
c_{ij} transportation cost between i and j;
E_{ij} externality measured in i and produced in j;
y_j 0/1 variable. It takes the value 1 when a facility is located in j, 0 otherwise;
x_{ij} proportion of demand located in i and allocated to j;
k number of facilities to be located.

Exact solutions are computed. When the externalities (E_{ij}) are set to zero, this model is equivalent to the *k*-median problem; hence, the *k*-median results are considered as the basis of comparison.

This model is particularly interesting for planning the location of (semi-) obnoxious facilities: consumers need these facilities and do not want to travel far (minimisation of the transportation costs), but they also do not want to be affected by pollution or other drawbacks. Hence, optimal locations of these facilities are a compromise between (1) the transportation costs minimisation that tends to locate the facilities centrally, and (2) the pollution minimisation that tends to locate facilities far from the consumers, at the outskirts of the studied area. Our interest is to see how far the optimal locations suggested by the accessibility criterion (*k*-median with no externality) are modified by the addition of a pollution criterion. What are the location changes? Do they occur continuously or stepwise? The conclusions of these simulations should be of great help for further operational applications: we know that important pollution leads to external locations (for example nuclear plants in France), but how sensitive are the model results to slight pollution levels? In our model, k – the number of facilities to be located – is exogenous. Given the size of the studied lattice of points, k will here vary from 1 to 9.

8.3 EXPERIMENTAL RESULTS

Computational results are now presented in order to show the sensitivity of the location modelling results to the importance of the negative externalities introduced in the objective function and associated to each edge of the network. Section 8.3.1 is devoted to the effects of changes in extension (*Ext*) and intensity (*Int*) on the average distance between the customers and the facility, as well as on the number of facilities k. Section 8.3.2 analyses the sensitivity of the optimal locations.

8.3.1 Externality and Optimal Average Distances

The average distance between demand points i and their closest facility j^* is a standard output in location-allocation practice, often considered as a measure of spatial efficiency; it is commonly used for comparisons between spatial solutions. In general, the average distance decreases when k increases. Moreover, the relative increase of the average accessibility is usually smaller for each additional facility. Figure 8.2 gives the k-median results computed for the transportation network, when no externality is considered ($E_{ij} = 0$). Let us analyse how far this trend is affected by the inclusion of externalities, how far the average accessibility of the services is modified by the inclusion of negative externalities.

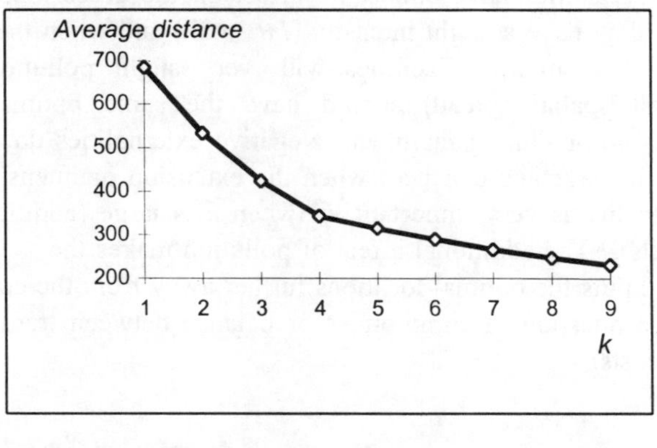

Figure 8.2 Expected decreasing relation between the average distance and the number k of optimally located facilities (k-median results)

Figures 8.3a to 8.3d illustrate the relative variation of the average distance with k (X-axis) for various values of *Ext* and *Int*. The average distance computed by the model for a given value of the externality (*Int/Ext*) is compared to the average distance provided by the k-median results computed with no externality; the comparison is expressed in percentages.

These percentages are reported on the Y-axis. Figure 8.3 is divided into four diagrams (noted 3a to 3d). Each one refers to a fixed value of *Ext* (300, 400, 500 and 600), and shows the variation with k ($k = 1$ to 9) of the relative difference between the average distances. Five values of *Int* (*Int* = 200, 300, 400, 500 and 600) are illustrated for each value of the extension. They were chosen among the large set of simulations undertaken. The results are representative of the observed trend.

In the studied 18 × 18 squared environment, locating few polluting facilities ($k \leq 4$) does not or little influence the efficiency measure, especially as far as pollution with restricted spatial spread (*Ext*) is concerned. When more facilities are located ($k \geq 5$), the effects of the changes in *Ext* and/or *Int* are greater and increase with k: optimal locations are modified when *Int* and/or *Ext* are very large. When k is small, transportation costs dictate the optimal solutions; when k increases the relative importance of the total transportation cost is reduced (a well-known phenomenon) compared to the total externality cost (this is a fairly obvious result). For sufficiently large values of k, externality costs start to matter and shifts in the optimal solutions are experienced. If the chosen maximum values of *Int* or *Ext* were larger (> 800), migrations of the optimal locations would appear for smaller values of k: the values of *Int* and/or *Ext* determine the location changes. If the transportation costs are very large compared to the pollution costs, they dictate the location mechanisms; if not, the polluting costs do.

Negative externalities have almost no effect on the accessibility of the services when they have a slight intensity (*Int* small) and when they do not spread much (*Ext* small). Facilities with very slight pollution (small intensity, small spatial spread) should have the same optimal spatial organisation as non-polluting facilities. Negative externalities do influence spatial optimality (average distance) when the extension or intensity of the negative externality is very important, or when k is large (addition of the pollution effects). The additional effect of pollution makes the system non-sustainable and puts the optimal locations further away from the consumers. Everything is a question of compromise or balance between transportation and pollution costs.

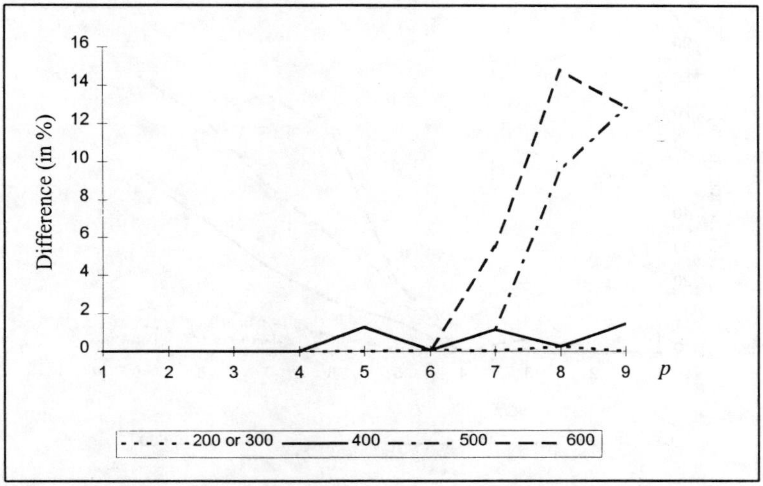

Figure 8.3a Variation of the average distance with the number of facilities and Int, Ext being fixed to 300

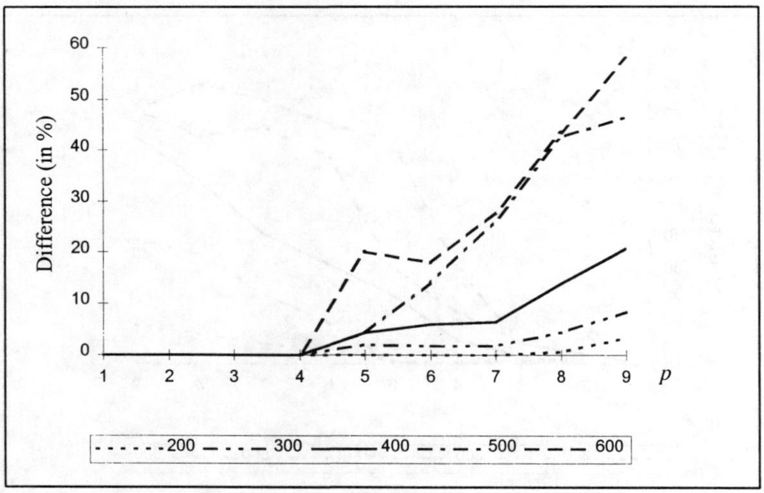

Figure 8.3b Variation of the average distance with the number of facilities and Int, Ext being fixed to 400

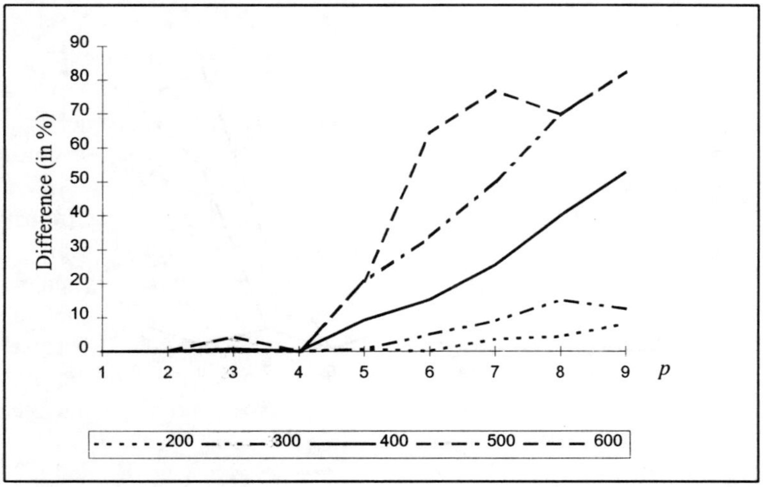

Figure 8.3c Variation of the average distance with the number of facilities and Int, Ext being fixed to 500

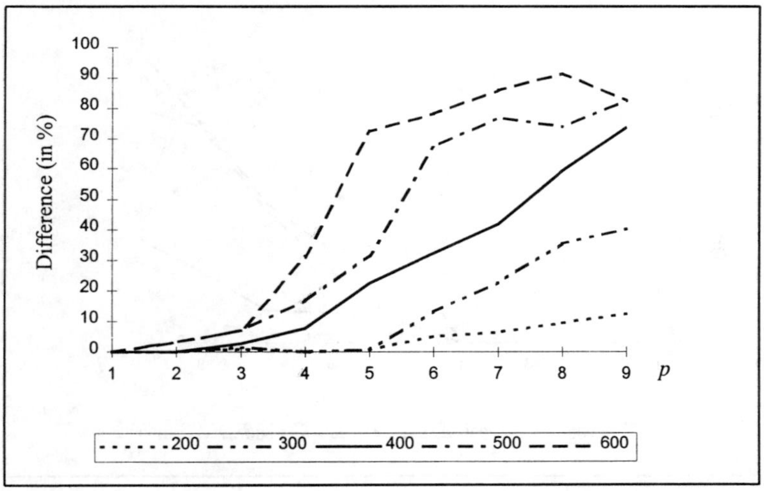

Figure 8.3d Variation of the average distance with the number of facilities and Int, Ext being fixed to 600

This process can easily be explained: transportation costs and pollution costs are here 'added' to each other, that is to say expressed in the same units.

If pollution is small compared to transportation costs, the quantity added to accessibility is very small and has little or no affect on total cost. When pollution increases, total cost increases, modifying the optimal locations and hence, the spatial efficiency of the planned service. Figures 8.4 illustrates this theoretically: transportation costs are supposed to increase linearly with distance from the facility when there is no pollution ($Int = 0$). Let us consider the total cost when a pollution is added to the accessibility measure. In Figure 8.4, the extension is fixed to 200 and the intensity varies from 0 to 800. When the intensity is small (compared to accessibility), the total cost function varies with distance from the facility. When the intensity increases, the variation of the total cost takes a maximum value at the locations of the facilities, modifying quite intensely the total cost variation with distance: as both costs are added, pollution becomes determinant. Hence, spatial efficiency is modified when the pollution is important compared to accessibility.

Let us now look at the sensitivity of the locations to the addition of externalities in a *k*-median model.

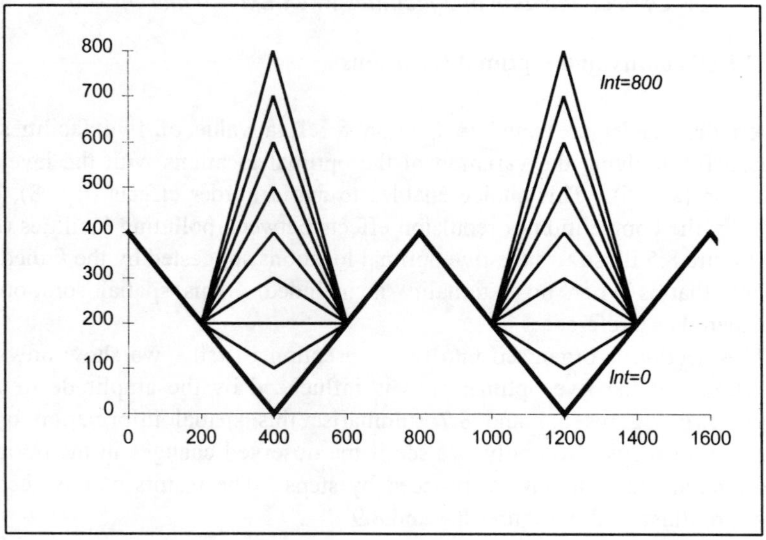

Figure 8.4 Variation of the expected total cost supported by a customer (transportation + externality) for Ext = 200

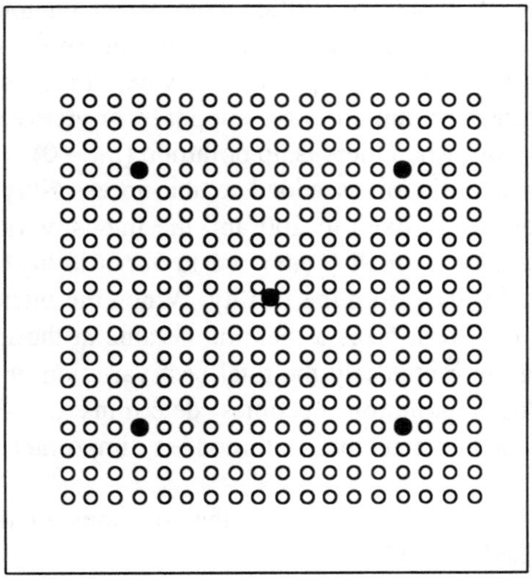

Figure 8.5 Optimal location of 5 facilities when Ext = 0 and Int = 0

8.3.2 Externality and Optimal Locations

Given the results presented in Section 8.3.1, a value of five facilities is chosen for studying the variation of the optimal locations with the level of pollution ($k = 5$). This choice enables to avoid border effects ($k < 8$), but includes the competition or repulsion effects between polluting facilities ($k \geq 4$). Figure 8.5 illustrates the five optimal locations suggested by the *k*-median model, that is when no externality is included. This spatial solution is considered as a reference.

This section is organised into two subsections: firstly, we show how far the choice of the five optimal sites is influenced by the amplitude of *Ext* and/or *Int*; Figures 8.6 and 8.7 summarise this spatial information by a selection of maps. Secondly, we see if the observed changes in the optimal locations are 'continuous' or proceed by steps. The results of this second 'test' are illustrated in Figures 8.8 and 8.9.

How far are optimal locations modified by the externalities?

Figure 8.6 gives the five optimal locations ($k = 5$), for a fixed value of the extension (*Ext* = 600) and for six selected values of the intensity (*Int* = 300 to 800 by 100). When the intensity is small (*Int* < 300), optimal locations are not influenced by the addition of externalities: the five optimal locations are those illustrated in Figure 8.5. When *Ext* increases (*Ext* ≥ 300), optimal locations are modified: the central point is first ejected from the centre of the lattice (for *Ext* = 300 to 500) and tries to find a place between the other four optimal sites. This centrifugal force pushes two locations to the border of the area (*Ext* = 500). When the intensity is very high (*Int* = 800), the system tends to put one site at each corner of the squared lattice and one in the centre; this configuration maximises the distance between the locations.

Figure 8.7 gives the optimal locations for a fixed value of the intensity (*Int* = 600) and for the six values of the extension (*Ext* = 300 to 800, in steps of 100). When the extension is small ($Ext \leq 300$), optimal locations are the same as those of the reference configuration (Figure 8.5). Starting with a spatial distribution of the supply sites that maximises accessibility (*Ext* = 300), we end once again (*Ext* = 800) with a spatial arrangement that maximises the distance between facilities. In between ($300 < Ext < 800$), the locations draw aside to avoid nuisance storage; some rotations of the spatial configuration appear in order to maximise the distance between supply sites. For each computed solution, the model finds a compromise between transportation costs and polluting costs.

Stepwise or continuous migrations?

Figures 8.6 and 8.7 are drawn as a cartoon: a picture is given for some selected values of *Int* or *Ext*. In this subsection, simulations were conducted with a much smaller step for the *Ext* and *Int* variations: the step is set to 25 instead of 100. In order to test if the migration process of the supply sites to the periphery is continuous or not, some complementary results are drawn in Figures 8.8 and 8.9. For a fixed value of the extension (*Ext* = 600), the same simulations were illustrated but this time with a step of 25 for the changes in *Int*. Figure 8.8 shows the point migrations for small critical values of *Int* (*Int* = 200 to 575); Figure 8.9 refers to larger values of *Int* (*Int* = 575 to 800). For small values of *Int* (Figure 8.8), the migration of the supply sites is almost continuous; for higher values of *Int* (Figure 8.9), changes are less coordinated. The locations draw aside in order to avoid nuisance storage; rotation of the geometric configuration also occurs in order to maximise distance between the optimal locations.

We can conclude that when pollution increases (in extension, or in intensity), the same spatial process occurs: first, expulsion of the centre which tries to find a location among the other 'competitive' points, and then, expulsion of four points out of five to the corners and reintegration of the central point. Below an intensity or an extension of 300, there is no effect of the externality on the optimal locations. Migration of the optimal locations occurs continuously at the beginning and are less coordinated for higher intensity of pollution.

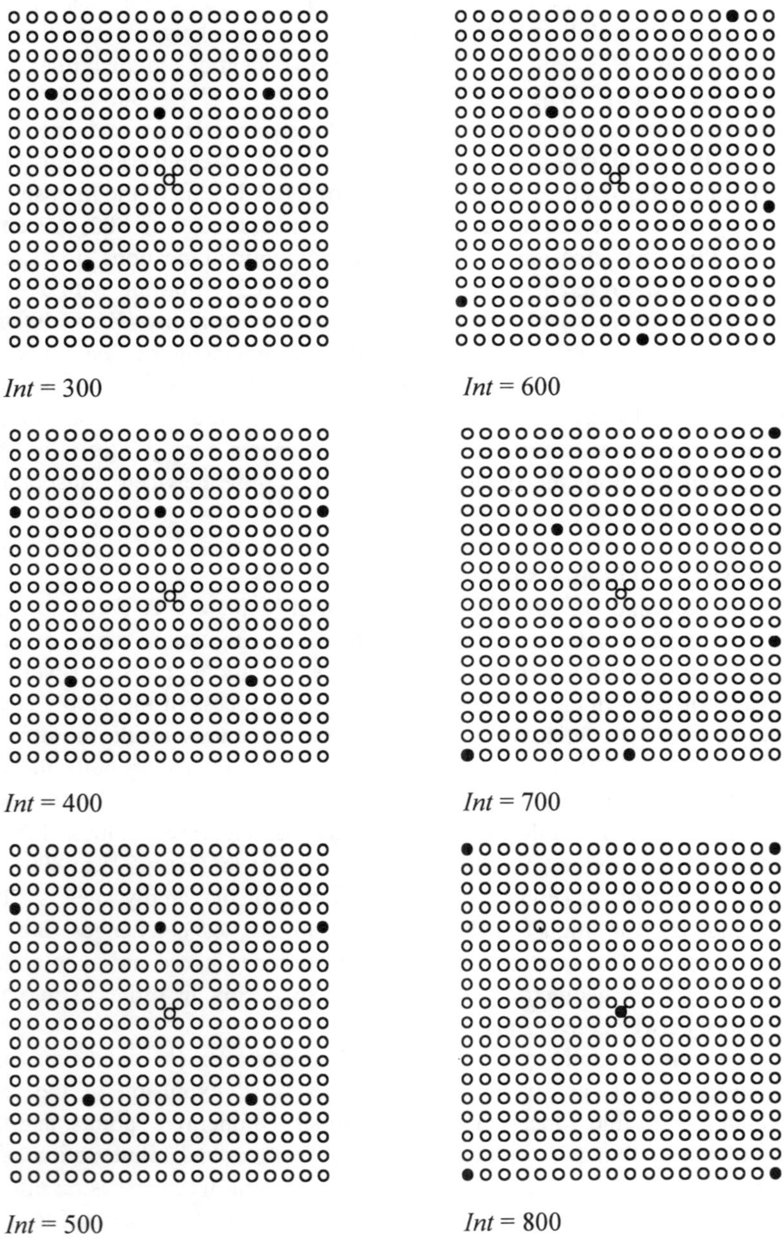

Figure 8.6 Changes with Int of the optimal locations (Ext = 600)

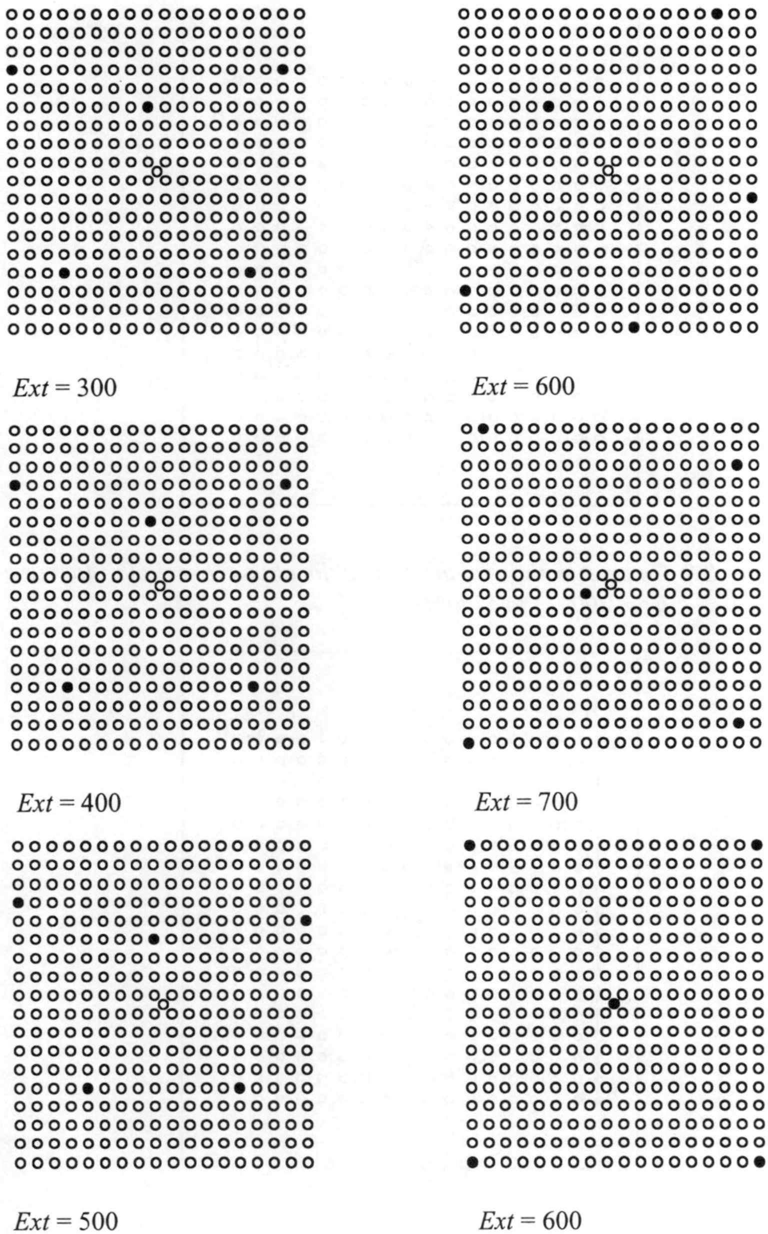

Figure 8.7 Changes with Ext of the optimal locations (Int = 600)

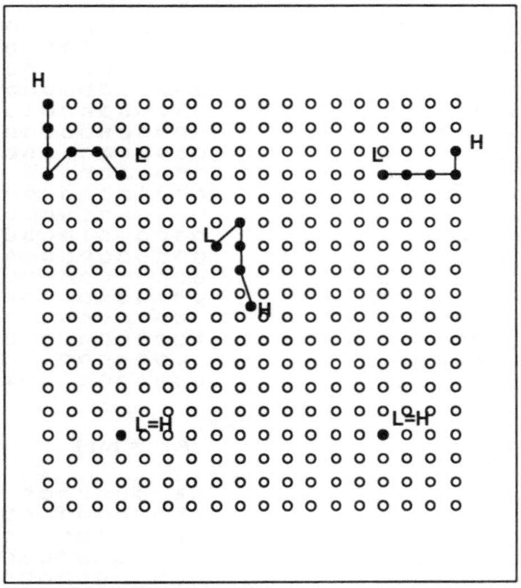

Figure 8.8 Changes with Int of the optimal locations (Ext = 600; Int varies from 200 (L) to 575 (H))

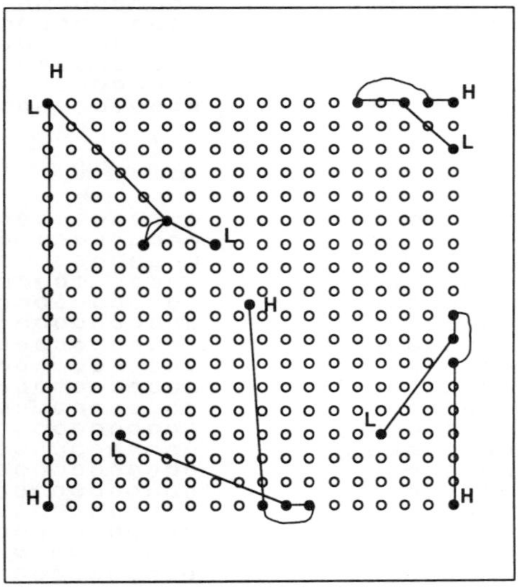

Figure 8.9 Changes with Int of the optimal locations (Ext = 600; Int varies from 600 (L) to 800 (H))

8.4 CONCLUSION

There are many real-world situations in which planners are faced with the problem of determining where to place a facility in a spatial environment (region, city, village) in order to minimise the undesirable effects of the new facility on existing population settlements and to minimise the total transportation costs. This chapter tests the sensitivity of the optimal location-allocation results to the introduction of negative externalities in the objective function. Externalities are included in the objective function of a modified *k*-median model, and simulations are performed on a regular squared lattice of 325 points.

Within the limits of the experiments, the performed simulations show that all location changes are generated by changes in the balance between transportation costs and polluting costs: slight pollution does not modify the solutions of the location-allocation model; high extension or intensity of the pollution generates 'competition' between facilities and modifies the reference locations. Facilities are located further apart from each other, and hence further from the consumers. Very high values of the intensity and/or the extension tend to push the optimal locations to the border of the studied environment. In other words, when the pollution is slight or medium, accessibility of the supply sites remains the most important location criterion. When pollution gets more important, facilities tend to be located as far as possible from each other, tending to another spatial optimum. In this latter case, transportation costs and hence the shape of the transportation network do not determine the optimal locations of the facilities. Applied work should be aware of this when finding the best spatial solution: pollution is not always a source of optimality changes.

In this chapter, we have addressed a practical well-known problem involving a local community, a region or a country that must build one or more undesirable facilities. Examples are numerous: nuclear power generating stations, hazardous material dumps, any potentially dangerous or undesirable facility such as a plant that has very noisy processes, a fire station or a hospital with a lot of emergencies and so on. One illustration is developed in Chapter 10. It concerns the optimal location of a community - recycling centre in a semi-rural Belgian township (La Bruyère). It confirms the results put forward in this chapter. Very often, local opposition is encountered when the location of such facilities is planned; this opposition is due to the disutilities that accompany these facilities. These disutilities may include health, safety or environmental risks, depreciation of land and property values, degradation of the quality of life and so on. In this chapter, we do not discuss the shape of the spread of the negative effects produced by obnoxious facilities. We just show how far the intensity of the negative externalities affects the optimal locations, how far the trade-off between transportation and pollution influences the location results. Hence, the weight associated to each transportation edge plays an important role in the location results especially when high spatial variations of the pollution costs are observed (trade-offs with transportation costs).

Scope for future work on this topic is still enormous: a multi-objective model should be used and its results compared to the one used in this chapter in order to achieve a better operationality of this kind of model; other lattices of points should be used as well as multiple transportation networks; and more complex shapes of the externality function should also be incorporated in order to make the problem more realistic.

As regards this research, we see that very large externalities annihilate transportation costs and lead to locations on the borders of the studied area. Hence, when transportation costs are not the unique criterion that is optimised, the structuring role of the transportation network is much less than the one observed and discussed in the preceding chapters.

PART FOUR

Land-use planning and the shape of the transportation network: two real-world examples

Introduction to Part IV

The purpose in this fourth and last part is to establish the plausibility of some of the results obtained on toy-networks in the preceding chapters, by looking at some real land-use problems. Toy-networks have enabled us to consider many sources of variation as constant. Real planning circumstances are multidimensional, and include many sources of variation that are often difficult to control for. Our intention is not here to alleviate all of them, but simply (1) to show by two simple examples how far the results obtained with real-world planning situations corroborate – or not – some of the preceding conclusions, and also (2) to illustrate how far the preceding results have real land-use planning meanings and consequences.

Hence, Chapter 9 considers the problem of locating primary health care facilities in a Third World country (one district of Niger, Africa). It shows how, in this type of region, accessibility and transportation structure highly depend on seasonal conditions: how far the rainy season disrupts transportation and travel for long periods. The *k*-median model is applied to two different transportation networks designed between a fixed set of demand points. Five planning scenarios are considered. The modelling results lead to quite important spatial differences and show that ignoring the rainy season can produce potentially disastrous results, while accommodating it can produce much better results. By this example, we confirm empirically the importance of the shape of the transportation network in location-allocation results. In countries affected by seasonal variations of the road conditions, it is quite important to evaluate and compare increase of efficiency due to the location of new facilities and due to investments in the transportation network. The conclusions presented in this chapter corroborate those discussed for toy-networks in Chapters 3 and 4.

Chapter 10 aims at applying the optimal location-allocating technique used in Chapter 8 (trade-off between transportation costs and pollution costs) in order to solve the problem of locating one semi-obnoxious facility in a residential area: the location of a community recycling centre in a Belgian commune (La Bruyère) is studied. A sensitivity analysis of the results to the importance of the nuisance gives more insights about the operationality of the method and also about the importance of the measure of the externalities (pollution) in such location-allocation problems. Conclusions corroborate the main results obtained with toy-networks in Chapter 8: only important negative externalities modify the location-allocation results obtained with transportation costs minimisation; slight or medium pollution does not modify the results proposed by the transportation costs minimisation.

9. Optimal locations of health centres in Niger: rainy season versus dry season accessibility

This chapter has two main objectives. Firstly, it aims at illustrating by a real land-use planning case-study the general problem discussed in this book: the influence of the shape of the transportation structure on the location-allocation results. A given regional planning problem is analysed for health care facilities under two different transportation realities: the rainy season and the dry season transportation networks. Several planning choices are made and evaluated in the light of the preceding results. Secondly, this chapter enables us to confirm the usefulness of location-allocation models when tackling the difficulties imposed by a Third World environment. The context of the application is the primary health care centres in the district of Filingué, Niger studied in a 'mémoire de licence' (Amoud, 1999).

9.1 INTRODUCTION

Used as spatial decision support tools, location-allocation models offer a good potential for enhancing public sector planning in developed as well as in developing countries (see Hodgson, Rosing and Shmulevitz, 1993; Peeters, Thisse and Thomas, 1998; Laurent and Thomas, 1999). Rural populations in less-developed countries are often faced to accessibility restrictions when trying to fulfil important needs such as getting to basic services (primary health care centres, schools, and so on). These restrictions have direct consequences on the quality of life and also structural, long-term consequences for the country's wealth (see for example Airey, 1991; Okafor, 1990; Tewari, 1992; Vasconcellos, 1997). Accessibility to basic services is often very poor. The precise extent to which development is impeded by this fact is largely unknown although there has been widespread speculation that improving access to important services would facilitate development (see for example Fisher and Rushton, 1979; Johnson, 1970; Tewari, 1992; Rietveld, 1993). Several factors impair accessibility in less-developed rural regions:

dispersed facilities that only offer limited support services, the lack of public transportation inside rural areas, and the social and economic conditions of those populations.

The Third World environment is different and often presents different problems when applying location-allocation techniques: differences in socio-economic structure, in transportation systems or in climatic conditions make the locations problems in those countries not always straightforward to solve (Clarke et al., 1996). Without careful study of the context of application, many impressive but less than useful applications may result (Rushton, 1984; Oppong, 1996). As mentioned for instance by Oppong (1996) or Rondinelli (1990), one of the most notorious problems encountered in applying models in those environments involves data limitations: fully operational integrated geographical information systems are rare, maps are frequently obsolete or non-existent, population data are usually outdated, unreliable, unavailable or not in a usable format often due to high spatial aggregation. Consequently, it is often difficult to apply location-allocation techniques and to determine if their results are reliable or not. This is however a common drawback of most planning models applied in rural areas of developing countries (Rietveld, 1993).

However, many developing countries are in the process of selecting new locations for providing basic human services to mainly rural populations. These location decisions are critically important to development because a large proportion of their population is rural and their transport and communication systems are often poorly developed and costly to use (Rushton, 1984). Hence, given the preceding chapters, our aim will here be to see how far the transportation system influences the optimal locations and allocations. In other words, does 'weighting the edges' influence the location results? In that sense, do transportation conditions during the rainy season lead to another definition of spatial optimality than the dry season conditions? What happens if the decision-maker decides to build or upgrade roads instead of multiplying health care facilities? This chapter will not relate to sensitivity analyses but will evaluate planning choices in the light of the results obtained in the preceding chapters.

The remainder of this chapter is organised as follows: the general and geographical contexts of the study are outlined in Section 9.2. Simple *k*-median analyses are then conducted and presented in Section 9.3; they are designed to accommodate the rainy season. Discussion of the results highlights the specificity of the Third World applications for location-allocation models but also confirms the importance of the shape of the transportation network in the location results.

9.2 BACKGROUND AND STUDY AREA

9.2.1 Background

Minimisation of distances to the facilities is an objective that is often chosen by location-allocation modellers; numerous examples can be found in industrial as well as in developing countries. The concern here is to locate primary health care facilities in a rural environment in Africa. Hence, maximising access for users is of prime importance (Vasconcellos, 1997). The relationship between distance and health care facilities is particularly critical in rural areas, where people have to walk long distances to widely dispersed facilities and must travel additional distances to nearby urban areas for higher-level facilities. With the lack of public transportation and an inability to pay for private transportation, most people do not benefit from a reasonable health programme. This distance-related problem concerns the analysis of accessibility – both in time and kilometres – as well as the issues of welfare and equity in access to basic public services. The access to primary health care centres is an important part of accessibility to basic activities, which have to be considered as a right in contemporary life. Hence, transport infrastructure and the availability of transport means can be considered as playing a vital role in attaining the basic needs of the rural population (Barwell, Edmonds and Howe, 1985; Begg, 1992; Nutley, 1984; Vasconcellos, 1997).

Locating healthcare facilities in order to maximise access for users in developing countries has already led to several interesting studies often bound to international contracts; unfortunately, publications (if any) are often to be found in quite 'confidential' journals (Rushton, 1984 and 1988). Let us talk about some exceptions and recall that Bennett, Eaton and Church (1982) applied a maximal covering approach to find locations for healthcare centres in Colombia. Mehretu, Wittick and Pigozzi (1983) used the *k*-median model with a maximal distance constraint to find locations of health service sites in Burkina Faso. Patel (1979) used a set-covering model for locating service centres in South Gujarat, India. Banreji and Fisher (1974) integrated a hierarchical location analysis in a set covering as well as in a *k*-median for area planning in rural India; Fisher and Rushton (1979) used the same methodology for India and Indonesia. Ayeni, Rushton and McNulty (1987) and Stock (1982) used location-allocation models to assess the improvements in accessibility resulting from the addition of new maternal, child welfare centres and health dispensaries in Nigeria. These authors concluded that simply increasing the number of facilities appears insufficient for improving health services in rural areas. Oppong and Hodgson (1994) and Oppong (1996) studied the configuration of health facilities in Ghana. They showed the usefulness and the limits of location models (set covering, *k*-median) in Third World settings and suggested several ways of accommodating the rainy season in the analyses. Oppong (1996) pointed out that those unique spatial circumstances present problems that if ignored could lead to seriously misleading conclusions.

In many Third World regions, the quality of roads varies from a bush track suitable for foot traffic only to all-weather roads for heavy vehicles. Many areas of the Third World are well beyond reach of a motorable track, seasonally or not. There is not only a lack of funds for road building, but also the physical difficulties of building and maintaining roads are enormous (Clarke et al., 1996, p. 241). Moreover, the rainy season often reduces travel opportunities considerably: in many regions, road conditions vary seasonally, some roads being not motorable in the rainy season, hence modifying travel patterns deeply.

9.2.2 Studied Area

The context of application here is the district of Filingué, which is one of the six districts of the department of Tillabéri in the western part of Niger (Africa). This district covers 24 558 km^2 and counted 286 451 inhabitants in 1988; the population density is very low (11.7 inhabitants /km^2) and for 96.8 per cent located in villages or hamlets of less than 3000 inhabitants. In the district, only three villages have more than 4000 inhabitants, the largest one being Filingué with a population evaluated at 9202 inhabitants. The economy is based on agriculture and stock farming. In our case, population of the hamlets has been aggregated to the closest administrative centre in order to handle a reasonable matrix of data; hence, 207 nodes are defined in order to represent the spatial pattern of demand (Figure 9.1). The weight attributed to each node is the total number of inhabitants in 1988, the latest census for which data were available for the present study, putting forward the acuity of the problem of data availability and quality (Section 9.1).

Climatically, the district is characterised by three seasons: one dry and hot from March to June, one warm and rainy from July to October and a third, dry and cooler, from November to February. The four rainy months represent an annual total of 321 mm of rainfall (Ministère du Plan, 1991). Based on the rainfall data alone, travel difficulties are only expected from July to October. It is however important to say that the seasonal effect in Filingué is not as acute as in other African regions where more important rainfalls are observed.

The transportation network in the district is quite sparse and poor, the south of the district having less kilometres of roads, but these latter are of better quality than in the north (Figure 9.2). Walking is the most popular mode of travel because few people own a motorbike or even a bicycle; moreover, commercial passenger vehicles charge high fares during the rainy season because the roads become virtually impassable.

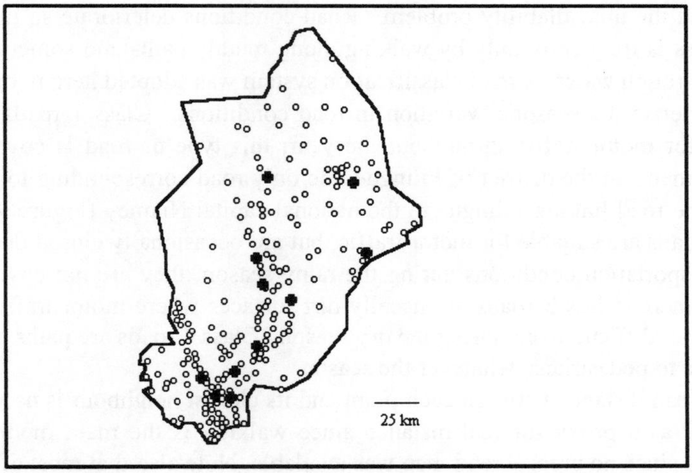

Figure 9.1 District of Filingué: 207 demand points (white dots) 11 existing facilities (black dots)

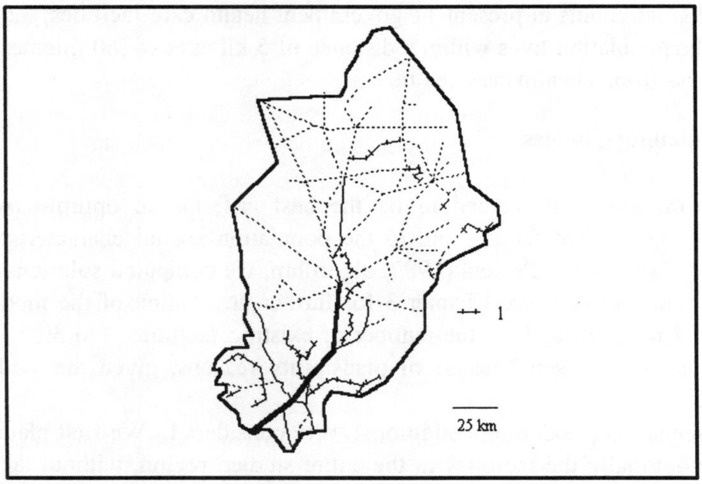

Note: 1 = suggested enhancements; dotted lines = class 3, full thin lines = class 2 and thick full lines = class1 roads

Figure 9.2 Suggested enhancements of the road system in Filingué

Inaccessibility is a perpetual problem that plagues the system of health care facilities in the Filingué district. During the dry season, when roads are generally motorable with little difficulty, availability of any transportation means is difficult. During the four months of the rainy season, bad roads compound the unavailability problem. Road conditions deteriorate so much that access is frequently only by walking along muddy paths and sometimes wading through water. A road classification system was adopted here in order to characterise the seasonal variation in road conditions. Class 1 roads are suitable for motor traffic throughout the year; this type of road is covered with bitumen. In the district of Filingué, the only road corresponding to this class is the road linking Filingué to the national capital Niamey (Figure 9.2). Class 2 roads are suitable for motor traffic, but are occasionally closed due to poor transportation conditions during the rainy season; they are not covered with bitumen. Class 3 roads are usually dirt surfaces where motor traffic is possible but difficult even during the dry season. Class 4 roads are paths only accessible to pedestrians, whatever the season.

Euclidean distance between each point and its closest neighbour is used in this study as a proxy for real distance since walking is the main mode of travel and since no precise road map was available. Note also that most of the boundaries of the district coincide with significant natural geographic barriers to movement, and that only 7 per cent of the villages are located along the class 1 road, accessible to cars all the year round.

The district counts at present 11 government health care facilities; 32 per cent of the population lives within a distance of 5 kilometres (60 minutes on foot) or less from a health care centre.

9.2.3 Modelling Choices

Distance minimisation seemed to be the best criterion to optimise with respect to the studied facility and to the population spatial characteristics. Using the Hanjoul and Peeters (1985) algorithm, we computed solutions for the *k*-median problem (see Chapter 3 for further description of the model); values of *k* range from 11 – the number of existing facilities – to 30. This latter value was chosen because of plausibility reasons, given the studied problem.

Five scenarios (modelling conditions) were considered. We first planned to locate optimally the facilities in the entire studied region, without taking into account the variations in the transportation conditions. Hence, all 207 nodes were considered as demand points as well as potential supply sites. Distances are the real distances measured in kilometres, whatever the type of road. Hence, given the economic and transportation conditions discussed earlier, we suppose that everyone goes on foot to the health care facilities (scenario 1). In the following sets of simulations, the spatial and seasonal variations of the transportation conditions are taken in account. There are several ways of accommodating them within the framework of a location-allocation model.

In scenario 2, facility locations are allowed only in places linked by a motorable road. Restricting the location of facilities on transportation axes will increase travel distances for the populations (comparison between scenarios 1 and 2) but permits better supervision and service, and thus offsets the logistical problems relating to seasonal influences. In this case, the analysis involves 112 nodes as potential supply centres; the distance is expressed in kilometres.

In scenarios 3 and 4, time distances are used as 'weights' to each segment of road instead of measured kilometres. Movements over class 2, 3 or 4 roads can then be given additional weight to reflect the difficulty of travelling along these roads during the rainy season. Hence, scenario 3 refers to the rainy season conditions only, while scenario 4 refers to the dry season conditions. Weights associated to the four classes of roads are reported in Table 9.1; they reflect the relative difficulty of travel and are simply a function of the average speed allowed on the roads. For Class 1 roads, we considered that the average speed is not affected by the seasonal conditions and that an average speed of 80 km/hour is sustainable. Hence, the associated weight is equal to 0.75 in both scenarios; it enables us to transform the kilometres used in scenario 1 into minutes of travel. In the two scenarios, 207 demand points and 112 potential supply nodes are taken into account.

Table 9.1 Road weights used in scenarios 3 and 4

Type of road	Dry season	Rainy season
Class 1	0.75	0.75
Class 2	1.00	1.50
Class 3	1.50	3.00
Class 4	12.00	18.00

Finally, scenario 5 considers an enhancement of the transportation network (this means that investment would first be done in constructing new transportation axes before creating medical care centres). The suggested enhancements are mapped in Figure 9.2; they mainly consist in joining villages that were presently not linked to the local transportation network by a class 3 road (a track). Following the definition of the set of potential supply sites used in scenario 2, 42 nodes were added to the 112 previous ones;

hence, $J = 154$ points. Weights associated to the distances are those defined for scenarios 3 and 4.

9.3 MODELLING RESULTS

The location analysis involves 207 demand points with the total population at each node considered as a proxy for demand. Distances are expressed in kilometres and are possibly weighted by a parameter reflecting the difficulty of travel; a different parameter is bound to each edge of the transportation network (scenarios 3 to 5). The number k of facilities to be located varies from 11 (number of existing facilities) to 30. Several k-median outputs are briefly analysed in order to evaluate the results of the five scenarios (mainly the average and the maximum distances, the optimal locations and allocations).

9.3.1 A Unique Season

Scenarios 1 and 2 refer to the distance expressed in kilometres; the transportation conditions are equal on all edges. Scenario 1 considers 207 potential supply sites, scenario 2 limits the number of potential sites to the 112 most accessible sites.

Let us first consider the current provision of services. When limiting ourselves to distances expressed in kilometres, the present (existing) system of 11 health care centres provides an average weighted distance of 12.2 km while an optimal system of 11 facilities chosen among 207 potential supply sites leads to a value of 10.7 km. Hence, the existing system involves 14 per cent more travelling than does the optimal one, which is quite reasonable and observed in many real planning circumstances. When considering 112 potential supply sites (scenario 2) the average distance is slightly higher (11.2 km).

Let us now add new facilities to the 11 existing ones. Hence, k varies from 12 to 30 facilities. The variation of the average weighted distance with k shows a regular decrease when k increases, very similar for scenarios 1 and 2: there is a very slight or nul difference when the number of potential supply sites is 207 ($J = 207$) (scenario 1) and when it is limited to 112 ($J = 112$) (scenario 2). Figure 9.3 refers to the variation of the maximum distance and leads to similar conclusions: differences in terms of average and maximum distances between scenarios 1 and 2 are small; they just slightly increase when k increases.

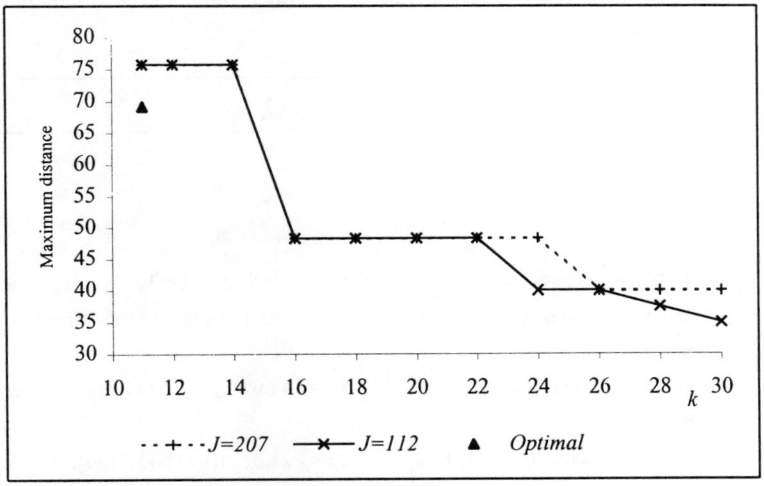

Figure 9.3 Variation with k of the maximum distance in scenarios 1 and 2

Comparing the 11 existing locations with the 11 optimal sites does not lead to dramatic differences: six sites out of 11 are chosen by the model among the existing facilities. When comparing the additional optimal locations issued from scenario 1 with those of scenario 2 (Table 9.2), the number of optimal sites to be different increases with k; the observed differences are however often spatially slight.

In scenarios 1 and 2, the transportation network remains the same and we assume that the road conditions are the same on all edges. Given these planning conditions, we can conclude that restricting the number of potential supply sites to more accessible nodes (scenario 2) has a limited impact on the number of facilities to be located. We observe some differences in the optimal locations, but they are slight: by definition, the model optimises accessibility. In the district of Filingué, the 112 potential supply sites are often located along the spinal roads which are also more accessible. Hence, in the case of Filingué, this has little impact on the final results.

Table 9.2 Index of the optimal sites for k varying from 12 to 30 in scenarios 1 and 2

k	Scenario 1	Scenario 2	D
12	87	87	0
14	9-87-109	7-87-109	1
16	9-38-109-111-201	10-55-109-111-201	2
18	9-38-84-109-114-156-201	10-55-84-109-114-156-201	2
20	9-38-44-84-109-114-147-156-201	10-55-84-92-109-114-149-156-201	4
22	7-13-38-44-84-95-110-114-147-156-201	7-13-55-84-92-109-114-149-156-186-206	6
24	7-13-38-44-84-92-103-114-121-147-156-186-206	7-13-55-84-92-103-114-121-149-156-168-186-206	3
26	7-13-38-44-84-92-103-114-121-137-147-156-168-186-206	7-13-29-55-84-92-103-114-121-137-149-156-168-186-206	3
28	7-13-28-38-44-83-84-92-103-114-121-137-147-156-168-186-206	7-13-29-55-67-84-92-103-114-121-137-149-156-169-186-197-201	7
30	7-13-28-38-44-83-84-92-103-114-121-137-147-156-168-173-187-192-206	7-13-29-55-67-84-92-103-114-121-137-149-156-169-173-189-192-197-206	7

Notes: index in bold refers to the differences

D = total number of sites that are different from one scenario to the other

9.3.2 A Rainy versus a Dry Season

In scenarios 3 and 4, we allow the use of several transportation modes and assume that the population will use a motor vehicle when possible and particularly when the roads are motorable. Hence, we consider the same physical transportation support as in Section 9.3.1 but we this time apply different weights to the edges in order to reflect the differences in transportation conditions during the rainy season (scenario 3) and during the dry season (scenario 4). Roads are divided into four classes of weights (Section 9.2.2); distances are now expressed in minutes.

Figure 9.4 represents the variation of the average weighted distance with the number of facilities. We always consider the location of the 11 existing facilities as given; hence, *k* varies from 11 to 30 (30 being the situation with the 11 existing facilities, plus 19 newly and optimally located ones). The figure also compares the case of the existing 11 facilities to the case of 11 optimally located facilities in both seasons (see Dry-optimal, Wet-optimal on the graph). 112 potential supply sites are considered. We can see that the average distance decreases regularly with *k*, in the same proportion whatever the season, but that travel costs associated with the rainy season are drastically higher (more than 50 per cent) than those associated with the dry season transportation network: 30 facilities located on the rainy season network are less efficient than 11 facilities located on the dry-season network. This confirms that additional facilities are not always required for accessibility to be improved (Oppong, 1996), as bad travel conditions influence optimal results.

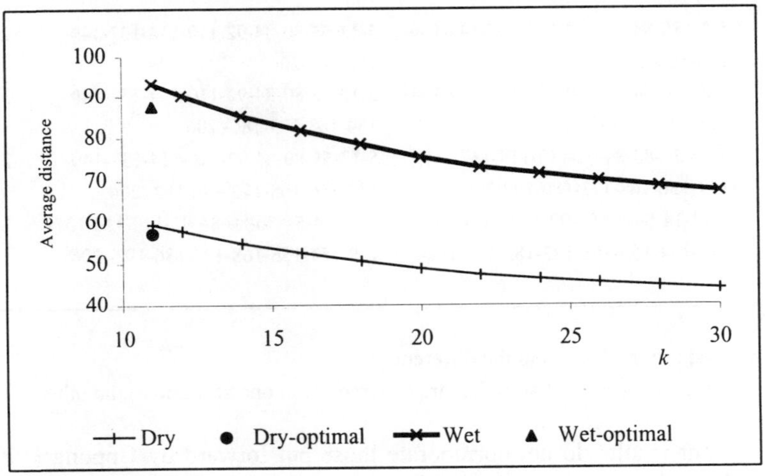

Figure 9.4 Variation with k of the average distance in scenarios 3 and 4

In the same way, maximum distances are much higher in the rainy season (588 minutes) than in the dry season conditions (392 minutes); these values are unaffected by the changes of *k*. These very high values are reached by the village Dané Fandou that is totally isolated whatever the season. This means that in the rainy season, this village is at best at 10 hours travel from any health care centre.

Table 9.3 and Figures 9.5 and 9.6 refers to the differences observed in terms of optimal locations for a fixed number of facilities; these two illustrations show that for a given value of *k*, there are changes in the chosen additional facilities for the rainy and the dry season, but at least half of the located facilities are common to scenarios 3 and 4.

Table 9.3 Index of the optimal sites for k varying from 12 to 30 in scenarios 3 and 4

k	Scenario 3	Scenario 4	D
12	**186**	**111**	1
14	111-137-186	111-137-186	0
16	5-**56**-111-137-186	5-**92**-111-137-186	1
18	5-**56**-92-111-137-**149**-186	5-**80**-92-**110**-111-137-186	2
20	5-**56**-80-92-111-137-149-186-**206**	5-**55**-80-92-**110**-111-137-149-186	2
22	5-**56**-80-92-110-111-137-149-168-186-206	5-**55**-80-92-110-111-137-149-168-186-206	1
24	5-**56**-80-84-92-110-114-137-149-**156**-168-186-206	5-**13**-**55**-80-84-92-110-114-137-149-168-186-206	2
26	5-**56**-80-84-92-110-114-137-149-**156**-168-173-186-192-206	5-**13**-**55**-80-84-92-110-114-137-**149**-**154**-168-173-186- 206	3
28	5-**14**-**56**-80-84-92-110-114-137-149-154-156-168-173-186-192-206	5-**13**-**55**-80-84-92-110-114-137-149-154-156-168-173-186-192- 206	2
30	5-13-14-**56**-80-84-92-110-114-137-149-154-156-168-173-185-186-192-206	5-13-14-**54**-56-80-84-92-110-114-137-149-154-156-168-173-186-192- 206	1

Notes: index in bold refers to the differences

D = total number of sites that are different from one scenario to the other

Hence, our results do not corroborate those put forward by Oppong (1996) who showed large operational differences between the rainy and the dry season. In our case, accessibility is much better during the dry season (and hence less facilities are necessary) but operational results in terms of locations are very similar when the same number of facilities is located. This however confirms the results obtained in Chapter 6 where we used several values of the parameters in the l_p-distances on a given transportation network: changing the weights associated to the edges does imply important differences in terms of absolute values of efficiency and/or equity, but leads to slight differences in terms of location results.

Hence, if the planner decides to build a given number *k* of facilities, there will be little or no change in the location choices between the two seasons. However, the measure of the average distance bound to the proposed solutions is much higher in the rainy season than in the dry season. The running and construction costs are the same, but the transportation costs are much different. Hence, in the dry season, less facilities are necessary in order to cover the region at a given level of efficiency. Accessibility matters.

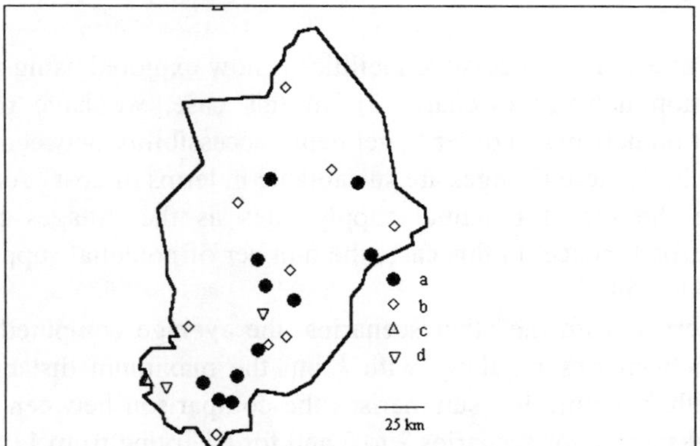

Note: a = 11 existing locations, b = optimal locations common to scenarios 3 and 4, c = sites chosen as optimal in the dry season only, d = sites chosen as optimal in the rainy season only

Figure 9.5 Optimal locations in scenarios 3 and 4 when k = 20

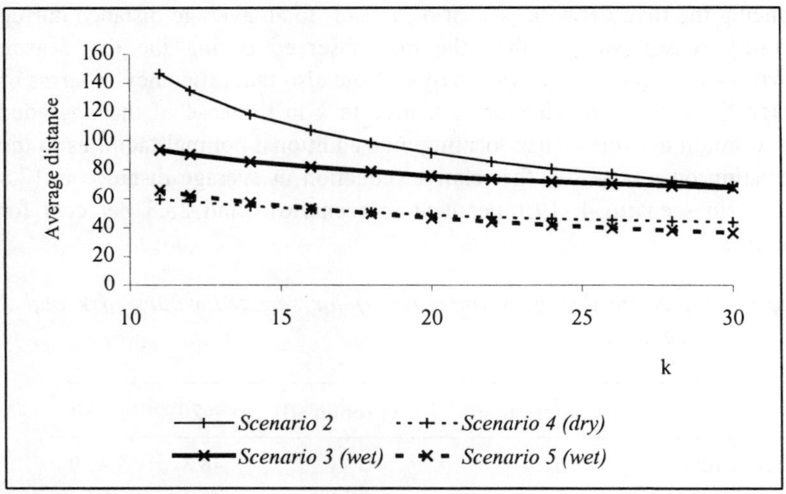

Figure 9.6 Variation with k of the average distance in scenarios 2 to 5

Now let us see what happens if the planner decides to affect part or most of his budget to improve the transportation network instead of multiplying the facilities (Section 9.3.3).

9.3.3 Rainy Season with an Enhanced Transportation System

Provision of efficient, accessible facilities is now explored using an enhanced transportation network (scenario 5). In this case, we have modified the transportation network in order to get better accessibility between the villages (Section 9.2.3); these changes are sustainable in terms of cost. As previously, we define the set of potential supply sites as the villages bound by a motorable road; hence, in this case, the number of potential supply centres is now equal to 154.

As observed with the other scenarios, the average computed distance in scenario 5 decreases regularly with *k* but the maximum distance does not change with *k*. Figure 9.6 summarises the comparison between the average weighted distances for scenarios 2 to 5 and for *k* varying from 11 to 30.

It shows that during the rainy season the 11 existing facilities on an upgraded road network (scenario 5) are as efficient in terms of average distance as 30 facilities on the existing road infrastructure (scenario 1). Table 9.4 gives some interesting values in order to complete the evaluation of the results: in terms of average distance, the 11 existing health care facilities linked together by a better road system (scenario 5) lead to almost the same average value (65.7 minutes) than 30 optimally located facilities during the rainy season with the existing road network (67.3 minutes). Moreover, enhancing the road network (scenario 5), leads to an average distance during the rainy season smaller than the one observed during the dry season (scenario 4) for much larger values of *k*. Note also that efficiency in terms of average distances is much more sensitive to *k* in the case of the upgraded transportation network: when locating nine additional optimal facilities to the 11 existing ones ($k = 20$), the relative reduction in average distance is 17.8 per cent in scenario 4, 19.1 per cent in scenario 3 and 28.8 per cent for scenario 5.

Table 9.4 Average distances (in minutes) for selected values of k and 3 scenarios

	$k = 11$ (existing)	$k = 11$ (optimal)	$k = 20$	$k = 30$
Scenario 4 (dry)	59.4	57.1	48.8	43.9
Scenario 3 (rainy)	93.2	87.5	75.4	67.3
Scenario 5 (rainy)	65.7	60.8	46.8	36.5

Table 9.5 refers to the indices of the selected sites for scenario 3 and 5 (rainy season on existing road network, rainy season on enhanced road system). Bold indices are those related to the sites that differ from one scenario to the

other. *D* represents the number of observed differences. We see that *D* increases with *k* and that the values of *D* are more important than those observed between dry and rainy seasons conditions (scenario 3 and 4, Table 9.3). Hence, in this case-study, enhancing the transportation network implies greater differences in the modelling results than just changing the weights associated to the edges in order to take the seasonal transportation conditions into account. Figure 9.7 and Table 9.5 illustrate the optimal locations for several values of *k*. In Figure 9.7, the 11 optimal locations suggested in scenario 3 (rainy season conditions on the existing road network) are compared to those proposed by scenario 5 (rainy season on the enhanced transportation network). Seven facilities out of 11 are common to both scenarios. The other four are quite different from each other. Globally, locations multiply in the south and are few and far between in the northern part of the district of Filingué. The more *k* increases, the larger the differences between the two scenarios. Hence, we see that the differences between scenarios 3 and 5 are spatially much higher that those observed between scenarios 3 and 4: changing the weights associated to the edges has less impact on the location results than adding new links.

Table 9.5 Index of the optimal sites for k varying from 12 to 30 in scenarios 3 and 5

k	Scenario 3	Scenario 5	D
12	**186**	**91**	1
14	111-**137**-186	**91**-111-186	1
16	**5-56**-111-**137**-186	**44-70-91**-111-186	3
18	5-**56-92**-111-137-**149**-186	5-**44-70-91**-111-137-186	3
20	5-**56-80-92**-111-137-**149**-186-206	5-**44-70-91**-111-137-**168**-186-206	4
22	5-**56-80-92-110**-111-137-**149**-168-186-206	5-**44-70-83-91**-111-137-**147**-168-186-206	5
24	5-**56-80-84-92-110-114**-137-**149-156**-168-**186**-206	5-**44-55-71-83-91-111**-137-**147**-168-**181-192**-206	9
26	5-**56-80-84-92-110-114**-137-**149-156**-168-**173-186**-192-206	5-**44-47-55-71-83-94-95-111**-137-**147**-168-**181**-192-206	10
28	5-**14-56-80-84-92-110-114**-137-**149-154-156**-168-173-186-192-206	5-**44-47-48-55-71-83-94-95-111**-137-**147**-168-173-**181**-192-206	11
30	5-**13-14-56-80-84-92-110-114**-137-**149-154**-156-168-173-**185-186**-192-206	5-**44-47-48-55-71-83-94-95-109-111**-137-**147**-156-168-173-**181**-192-206	12

Notes: index in bold refers to the differences

D = total number of sites that are different from one scenario to the other

Figure 9.8 is the last figure illustrating the differences observed between the results provided by the five scenarios used in this example. It gives the variation with the number *k* of facilities of the ratio of the maximum allocation to the minimum allocation; this ratio is a measure of the hierarchy within the located facilities. On average, hierarchy increases when the number of facilities increases but it particularly increases for scenarios 3 and 4 compared to 2 and 5. Enhancing the road network will lead to a better spatially balanced configuration.

At least three important conclusions can be drawn from the above scenarios and illustrations. Firstly, the *k*-median results are sensitive to the shape of the transportation network. Hence, we confirm the results obtained earlier by Oppong (1996, p. 135) who states that 'extra travel cost required to ensure year-round accessibility is low compared with that from an optimal system in which seasonal inaccessibility is ignored'. Secondly, reshaping the transportation infrastructure (scenario 5) has more impact on the location-allocation results than just reweighting the edges of the graph (scenarios 3 and 4). This means that changing the weights associated to the edges (Chapter 6) has less impact than modifying the transportation network (Chapters 3 and 4). Thirdly, the problem of accessibility to basic services in Third World countries is not to be solved by a simple multiplication of the service centres, but also by rethinking the general transportation problem (Vasconcellos, 1997).

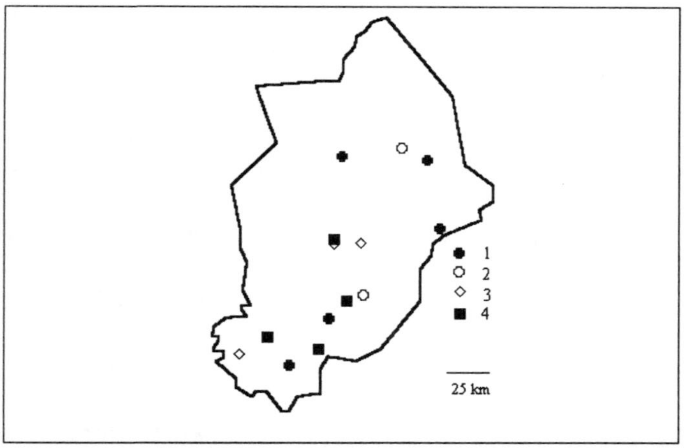

Note: 1 = existing locations selected by scenarios 3 and 5, 2 = new optimal site selected by both scenarios, 3 = optimal location for scenario 3 only, 4 = optimal locations in scenario 5 only

Figure 9.7 Optimal locations in scenarios 3 and 5 when k = 11

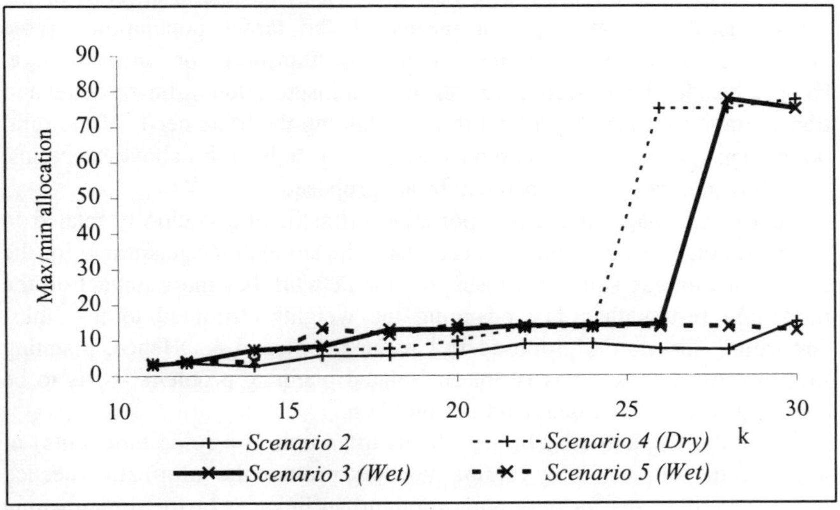

Figure 9.8 Hierarchies generated under the 4 main scenarios

9.4 CONCLUSION

This chapter illustrates with a real-world example the influence of the shape of the transportation network in location-allocation results; it also shows the need to be cautious when applying location-allocation models in Third World countries or to any other transportation circumstances where temporal variations in the transportation conditions exist (such as congestion). This chapter also confirms the sensitivity of the location results to the inputs of the model in general, and more particularly to the transportation costs. In this example, the rainy season limits the accessibility of the medical amenities so that (1) ignoring it would lead to fallacious operational results, and this more in terms of efficiency and equity measures than in terms of locations, and (2) upgrading the transportation system (and the transportation opportunities) may be more efficient than just multiplying the number of facilities.

Rural populations in developing countries face severe accessibility restrictions because of their geographical isolation, the commonly low level of transport supply and their socio-economic conditions. This distance-related problem concerns the analysis of accessibility as well as the issues of welfare and equity in the access to public services. Health care facilities in the district of Filingué – and indeed in most developing regions – would clearly be more successful if a shift was made from locating medical facilities on the basis of an administrative network to a more careful analysis of the actual distribution and specific needs of the target population; viable alternatives here include improvements in transportation infrastructure. Hence, beside the provision in services, transportation infrastructure and transportation means play a vital role in attaining the basic needs of the rural population (Nutley, 1984; Vasconcellos, 1997). Given the above scenarios, the following major recommendations are proposed.

Firstly, the shape of the transportation infrastructure is closely related to the operational location-allocation results. The simulations performed for the district of Filingué show that reshaping the network has more impact on the modelling results than just changing the weights attributed to the links, confirming the results proposed in Chapters 3, 4 and 6. Hence, planning locations of vital facilities is not an isolated planning problem; it has to be studied as a global regional land-use problem.

Secondly, location-allocation methods often optimise some function(s) of access to the facilities; they cannot, however, provide the information needed by regional planners for improving rural–urban linkages or for strengthening the function of economic growth centres in rural regions. With some tentative exceptions, location-allocation techniques do not simultaneously optimise transportation and services infrastructure (Rondinelli, 1990; Tewari, 1990). Hence, it is necessary to develop further methodologies for formally integrating location-allocation analyses with other forms of regional analyses that are being used in support of the development effort. This is the challenge of 'intelligent GIS' (Birkin et al., 1996).

Thirdly, it seems necessary to search for quantitative or statistical measurements of the differences observed between simulation conditions. How significant is an observed difference in location efficiency? Moreover, if location is an advantage, can it be used to counteract significantly other disadvantages in particular areas? Statistical measures of the effects attributable to improved access are needed .

Fourthly and last, another interesting further research programme would be to add to the location results a cost–benefit analysis between the costs associated with the enhancement of the transportation network and the benefits associated with the reduction in transportation time.

10. Optimal location of a recycling centre in Belgium: externalities versus transportation costs

This chapter aims at illustrating the results obtained in Chapter 8: with a real world land-use planning problem, it corroborates the results obtained previously on theoretical squares and hexagons. The studied example consists of optimally locating a community recycling centre in a Belgian residential commune (La Bruyère) in order to minimise the transportation costs and also to minimise the negative externalities undergone by the local population. In other words, each demand node is characterised by its position on the transportation network (transportation costs) and also by its location in the polluted space (pollution costs). The optimal location is a compromise between both types of costs. The essence of this chapter comes from a 'mémoire de licence' (Flahaut, 1997) directed by the author and also published in a paper (Flahaut, Laurent and Thomas, 2001).

The remainder of this chapter is organised as follows. In Section 10.1, the problem of locating a garbage sorting centre is discussed as a whole; in Section 10.2, the externality is described as well as the design of the experiment, the inputs of the simulation programs and the location model. In Section 10.3, the results of the application and the simulations are presented in terms of efficiency and optimal locations. Finally, Section 10.4 is devoted to conclusions and possible extensions.

10.1 INTRODUCTION

Managing garbage is a great problem for modern societies today and it will be certainly be so in the next decades. Numerous aspects have to be considered among which some spatial components that are worth studying by spatial scientists. In a country like Belgium, garbage is collected by trucks from households: several rounds are organised, depending upon the nature of the waste (paper, plastic, and so on). The frequency of the rounds depends on the degree of urbanisation of the area. In many 'communes', residents also have the opportunity to go by their own means to a community recycling centre (called *parc à conteneurs* in Belgium, and *déchetterie* in France). A community-recycling centre is a plot of land on which several containers are available, each assigned to one type of waste (glass, paper, cardboard, iron and so on). Hence, it is a place where residents sort their garbage or bring their cumbersome waste. In Belgium, at the time of the study, the use of a local recycling centre is often restricted to the inhabitants of the township. Constructed recently, these centres are built in such a way that underground pollution is almost non-existent. Most inhabitants want such a centre in their area and do not want to travel too far away to use it, but no one wants it next to his door: noise, smell, visual pollution and traffic are often invoked as perceived sources of nuisance. The will to locate this kind of facility 'close, but not too close' is popularised by the name of the NIMBY phenomenon (Not In My Back Yard). This makes the location problem interesting for planners and decision-makers. Hence, this chapter aims at suggesting a simple operational method in order to find the optimal location of a recycling centre and also at showing how far the results of the used location-allocation model are sensitive to the importance of the nuisances. Simulations are conducted on a real-world case-study: a Belgian residential 'commune' where such a location decision had to be taken by the local authorities in 1996.

As already mentioned in earlier in this book, location-allocation models often optimise a social welfare criterion such as the total transportation costs minimisation; distance is then often a physical distance or a transportation cost measured on a transportation network. The introduction of externalities in location-allocation models implies changes in the models' formulations as well as in the resolution methods: in addition to transportation costs, the model includes a minimisation–maximisation function of the cost–benefit of externalities on the nearby environment (see Chapter 8). Hence, in the case of a facility that generates pollution, the values associated with the edges of the transportation network are no longer restricted to a physical distance, but they are assumed to correspond to a function of some transportation costs and some pollution costs. Both costs (transportation and pollution) are unwanted by the users of the facilities: they want a facility to be as close as possible and they also want the negative effects they have to undergo to be as small as possible. The issues of accessibility to facility services and of minimising exposure to nuisance of communities provide an interesting planning situation in that competing siting objectives are involved.

Let us here recall that the location of obnoxious facilities has raised many fundamental problems in the last few decades. The majority of the authors use a model that maximises some function of distance, because their priority is the minimisation of the obnoxious effect of the facilities (see Chapter 8 for a review). Genarally speaking, the number of criteria to be optimised in the models found in the existing literature is variable: two-objective (for example ReVelle et al., 1991; Stowers and Palekar, 1993), three-objective (for example List and Mirchandani, 1991; Erkut and Verter, 1995) or even five-objective formulations can be mentioned (Current and Ratick, 1995). Erkut and Neuman (1989) presented a fairly comprehensive review of the existing literature about undesirable facilities in which they established a taxonomy of maximisation objectives. They underlined that little research has been done on multi-objective location problems and they proposed in 1992 one approaching this. Their model is a three-objective mixed integer programming formulation, where the objectives are to minimise total cost, minimise total opposition and maximise equity.

Valea, Baetz and Tsanis (1998) recently developed a simple model for locating recycling depots in a community of 22 000 inhabitants; their model is however limited to the maximisation of the coverage of a depot site with constraints based on projected 'recycler behaviour'. The model used in this chapter intends to achieve simultaneously the two following objectives: (1) maximise the accessibility to the facilities, and (2) minimise the obnoxious effect of these facilities.

10.2 THE MODEL AND ITS INPUTS

10.2.1 The Model

The aim is not here to incorporate the multiple trade-offs and conflicting concerns which arise when locating undesirable facilities. Hence, our model is biobjective, including only the trade-off between two conflicting concerns that are the (1) location close to an inhabited area for the reason of transport costs minimisation, and the (2) location far away from the population in order to limit exposure to nuisances. These two objectives are clearly in opposition in a well-delimited studied environment, but both have to be considered in modelling optimal location problems of semi-obnoxious facilities. More precisely, we aim at testing the effects of the introduction of negative externalities in a location-allocation model on the operational results, externalities being fixed costs to be added to the transportation costs. This addition may be seen as rather simplistic: other authors have solved this modelling problem by multicriteria models minimising transportation costs and externalities (Ratick and White, 1988; Erkut and Neuman, 1992; Murray et al., 1998). It has already been shown (Hansen, Peeters and Thisse, 1981) that our approach is a special case of the bicriteria approach where the conversion factor between transportation cost and pollution objectives is assumed to be known. This explains why we had to calculate pollution costs (Section 10.2.2) that can be added to transport costs: both are expressed in the same unit (monetary unit). Moreover, our aim is also to show the sensitivity of the optimal location to increasing levels of nuisance and to show the operationality of the model.

We here consider a discrete formulation: facilities can be established only among a finite subset of nodes. A simple location-allocation model is used (the *k*-median) and modified in order to include externalities (see formulation in Chapter 8): the model minimises the sum of the transportation costs and the external costs characterising the impact of a facility. This problem can be seen as an extension of the simple plant location problem containing a cardinality constraint. Exact solutions are computed. When the external costs (E_{ij}) are set to zero, this model is equivalent to the *k*-median problem.

Our interest is now to see how far the optimal locations suggested by the accessibility criterion (*k*-median with no externality) are modified by the addition of a pollution criterion. What are the location changes? How far are the results sensitive to the importance of the externality? The conclusions of these experiments should be of great help for further operational applications: we know that important pollution leads to external locations (for example nuclear plants in France), but how sensitive are the model results to slighter pollution levels? Moreover, if the externality is only roughly or approximately measured, does this matter for the optimal location results and what impact does it have on the consequent operational decisions?

10.2.2 The Inputs

An undesirable facility generates a nuisance (a negative externality) diffusing in the surrounding environment (see Chapter 8, or Smith, 1977). Various types of impacts produce different proximity-sensitive zones: for example, noise emanating from a local recycling centre tends to dissipate more quickly with distance than the air pollution impacts of a heavy industry. Most externalities do, in fact, have a fairly localised impact and their effects are weakened by distance from the source (Buchanan and Stubblebine, 1962; Mishan, 1971; Chapman and Walker, 1987; or Verhoef, 1994). Harvey (1971), Smith (1977) and Bale (1978) visualise this in terms of a spatial externality field, with a downward sloping externality gradient in all directions from the source. If pollution decreases linearly with distance, the externality field can be depicted in terms of a cone, where an even population distribution surrounds the source or a set of contours. The intensity is here defined as the quantity of pollution measured at its source; the spatial extension of the nuisance is the maximum distance from the source at which the pollution is identified (see Figure 8.1).

Several types of nuisances can potentially be associated with a community recycling centre. However, the noise generated by the loading and the removal of the filled containers by trucks seems to be the most important one (demonstrated by local surveys, Bureau Economique de la Province de Namur, 1996) and will be the only one considered here. Nuisances resulting from additional private road traffic would also be interesting to take into consideration, but modelling nuisances emitted by mobile sources come under a completely different methodology. Hence, a community recycling centre is here considered as a nodal source of noise.

The noise produced by the container removals is estimated as follows. Each movement of a container is considered as an individual event, and the number of containers that have to be removed by lorries depends upon the number of visitors to the recycling centre and on the quantity of waste produced. Negative externalities associated with the recycling centre are then the total noise generated by the truck movements on a yearly basis.

The introduction of externalities in location-allocation models requires the conversion of externalities and transportation costs into the same units of measure. The externality is by definition a non-marketable concept and the estimation of its monetary value is not easy. Among the techniques of direct valuation, the hedonist prices method and the contingent valuation are the most usual practices to elicit information about households' preferences for many environmental goods and services. The hedonist prices method seems appropriate to convert a noisy pollution in a monetary value by the way of the housing values. The methodology is to estimate a semi-logarithmic hedonist price equation where the logarithm of property price is regressed upon a range of variables. Two groups of regressors are introduced in the equation: the first contains the internal attributes of individual properties, and the second includes the attributes that influence property price through externality effect outside the control of individual owners or tenants. The second group of variables captures the influence of 'neighbourhood quality'. The hedonist price method is often used to determine the impact of environmental factors on the property market (see Soguel, 1994; Mayeres et al., 1996; Tomkins et al., 1998). Soguel (1994) showed that in American and European studies, we can observe an average individual's willingness-to-pay (*WTP*) of 1 per cent of the rent of the house to reduce the noise level by 1dB: it is the cost of the noise nuisance, that is called an 'external cost' because it is the cost of an externality. Following Soguel (1994), we here consider that 0.9 per cent of the house rent can be spent by users to reduce the noise level by 1 dB. Hence by using this value of 0.9 per cent , we do not intend here to evaluate precisely the individuals' *WTP*, but we try to examine which values of external costs are susceptible to generate modifications in the optimal locations of facilities, within a range of realistic levels of external costs. This approach enables us to make out a general outline within which it will be easy to place later exact and reliable estimations of these external costs.

This estimation is related to a continuous noise level during a period of time and not successive noisy events, as it is the case here. Therefore, we have to calculate an equivalent continuous noise level (L_{eq}) that contains the same total energy as several noisy events during a given period of time. The relation is established with the following formulation (Liénard, 1978):

$$L_{eq} = 10 \log \frac{1}{T} \int 10^{L/T} dt$$

where T is the period of time of observation and L is the measured level of noise. The L_{eq} is a continuous measure of noise level expressed in dB (A); a monetary equivalent can then be computed by hedonist methods. These estimations of the external costs of noise can then be introduced in the location model.

Modelling pollution spread has led to a lot of research, because pollution does not often diffuse just along road networks: air, water and soil diffusions are concerned and interfere with each other. Authors have tried to fully represent the impact of pollution on the environment and have elaborated some complex but realistic models (see for example Karkazis and Boffey, 1994; Matzoros and Van Vliet, 1992; Papadimitriou and Karkazis, 1991). In many instances, nuisance decreases nonlinearly with distance: Melachrinoudis and Cullinane (1986a) assumed, for example, that nuisance is inversely proportional to the squared distance, whereas Erkut and Neuman (1992) experimented with a general nuisance function where distance was raised to some negative power.

The externality considered here – the noise – decreases non-linearly with distance from the polluting facility. More precisely, the decrease is inversely proportional to the squared distance whatever the direction; we also consider that the atmosphere does not interfere. The function used in this study is the following (Flahaut, 1997):

$$f(d) = \frac{A}{B + Cd^2}$$

where

$f(d)$ = yearly external costs for a household located at a distance d from the facility;

d = distance to the polluting facility;

A = yearly external cost for a household located at a distance 0 from the facility;

B = the existence of *B* causes a small 'translation' of the decreasing curve, in order to avoid an indetermination of the function when d = 0. Its value is fixed to 1.0;

C = enables us to modify the slope of the decreasing function. *C* represents the friction of distance. Three values are used to represent 3 sizes of the dispersion area ($C = 10^{-4}, 10^{-5}, 10^{-6}$).

10.2.3 The Studied Environment

The Belgian 'commune' of La Bruyère is used as a support for the application. La Bruyère is a semi-rural residential township located 10 km north of Namur (Wallonia) and 40 km south of Brussels (Belgium). This commune covers 53.02 km^2 and counts 7138 inhabitants in the census of 1991. It is made of seven villages, each of them consisting of a dense centre and several peripheral hamlets. Hence, the resulting landscape is characterised by discontinuities in the habitat notwithstanding a recent phenomenon of rurbanisation and suburbanisation. The commune is located in a rich traditional agricultural region: large fields of beetroot, corn and wheat occupy the agricultural land still available between the hamlets.

A discrete formulation is much more convenient for land-use planning purposes in location-allocation sutides. Hence, the problem consists now in representing the sutided area by a discrete set of nodes and edges. The township of La Bruyère is officially divided into 25 statistical sectors. As those wards were defined in the early 1970s (Brulard and Van der Haegen, 1972), some of them had to be subdivided in order to better reflect the actual geographical disparities and better achieve our objective – similar size and shape, minimisation of population heterogeneity (Flahaut, 1997, p. 41). Hence, the commune was divided into 52 spatial units that maximise internal homogeneity (see Figure 10.1). Each spatial unit is represented by a point located at the centre of the built surface; all nodes are considered as demand points as well as potential supply sites. A weight is associated to each point; it represents demand and is here equal to the total population. All nodes are related by edges to their closest neighbours; edges represent the transportation network. Distances are measured in kilometres along the existing roads and associated with the edges; given the small differences in the speed conditions on the roads of the commune, no speed differential is added. Distance represents the disutility of the people when using the community recycling centre; the transportation costs are proportional of the number kilometres and a unitary cost of transport of 12 BEF/km (0.3 EUR/km) is adopted (Flahaut, 1997, pp. 53–4).

10.3 MODELLING RESULTS

Several simulations were undertaken in order to see how far the location results are sensitive to the changes in nuisance levels, these latter being determined by the pollution intensity and by its spatial extension. In Section 10.3.1, we refer to a realistic land-use planning situation where only one community recycling centre is to be located. In Section 10.3.2, the experiments are extended to three supply sites and to other measures of demand.

10.3.1 The Location of One Community Recycling Centre

The objective function of the model used here is the minimisation of the sum of the transport costs for the use of the service and the pollution costs (see Chapter 8). When pollution costs are set equal to zero, the model corresponds to the simple *k*-median problem. Hence, the results of the *k*-median model can here be seen as a benchmark. In the case of La Bruyère, the optimal location is node 15 (Figure 10.1), that is to say the centre of the most important village of the township (Rhisnes). This is not a surprise: it is the most populated area that also corresponds to the location of most local administrative services of the 'commune' (police and so on).

Let us now increase the levels of nuisance emanating from the recycling centre. Such levels can be obtained by increasing the 'intensity' of the nuisance, the spatial 'extension' of diffusion, or both. Results are summarised in Table 10.1 and Figure 10.1.

Any intensity of nuisance (expressed by a willingness-to-pay *WTP*, the maximum being equal to 79 780 BEF at a distance $d = 0$ from the source of pollution) associated with a short diffusion (expressed by a high friction of distance with $C = 10^{-4}$) has no effect on the optimal location. Hence, accessibility is the decisive criterion to determine optimal location of facilities with a low dispersion of emitted pollution. Note that with 'intensity' of nuisance, we do not consider here the intensity of the noise in itself, but the level of external costs we have to compare to transport costs associated with one visit to the recycling centre. The difference is that the frequency of visits by users to the recycling centre plays an important role in the evaluation of these external costs. With the same noise intensity in a given period of time (in this case one year), external costs will take proportionally a growing importance with regard to transport costs when the number of visits to the service becomes smaller.

Table 10.1 Main results for k = 1

WTP (in BEF)	Diffusion factor *C*	Optimal location (1)	Average distance (2)	(3)
≤ 79 780	10^{-4}	15	3600	0
≤ 31 720	10^{-5}	15	3600	0
≤ 9887	10^{-6}	15	3600	0
≥ 42 293 and ≤ 47 740	10^{-5}	11	4100	+ 14
≥10 711 and ≤ 16 741	10^{-6}	11	4100	+ 14
58 313	10^{-5}	9	5218	+ 45
≥ 63 760 and ≤ 79 780	10^{-5}	38	6017	+ 67
≥ 19 304 and ≤ 31 720	10^{-6}	38	6017	+ 67
≥ 42 293 and ≤79 780	10^{-6}	37	6388	+ 77

Note: (1) = index of the node
(2) = single way, in metres
(3) = Mean distance variation related to the optimal location for the *k*-median (spatial unit n°15), expressed in percentages.

Source: Flahaut (1997)

When the spread of pollution increases ($C = 10^{-5}$ and 10^{-6}), total external costs undergone by the close population are higher and the remoteness criterion enters in to the location process. Then, the optimal location is further away from the centre because external costs decrease with distance more than transportation costs increase, resulting in a global decrease of both costs.

With the highest realistic levels of nuisances considered in this work (*WTP* varying from 63 760 BEF to 79 780 BEF with $C = 10^{-5}$ and from 19 304 BEF to 79 780 BEF with $C = 10^{-6}$), the optimal location is more distant from most inhabitants: it is located at the northern upper limit of the 'commune' (node 37 or 38 on Figure 10.1), a much less densely populated area.

These results confirm those obtained on toy-networks (Chapter 8): a low level of externalities does not modify optimal locations suggested by the *k*-median problem; when the level of externalities increases, the location becomes progressively peripheral.

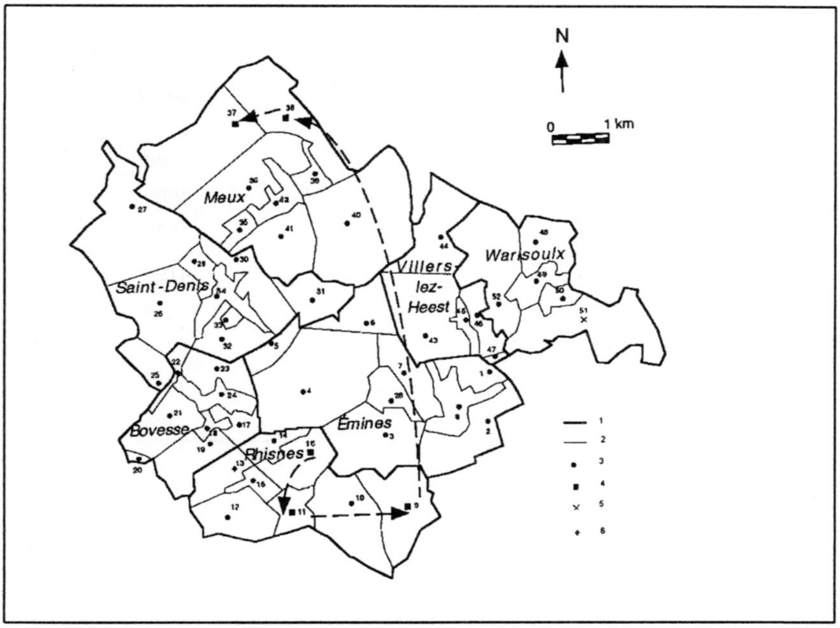

Notes: (1): limit of the 'communes'
(2): limit of the statistical wards
(3): demand node
(4): optimal site
(5): site corresponding to the maximum of the transportation costs
(6): current location.

Source: Flahaut (1997)

Figure 10.1 Optimal location of one recycling centre (k = 1) for increasing values of negative externalities

These results are in agreement with the semi-obnoxious nature of the public service considered in this study. Indeed, the most distant node from the population (node 51, associated with a mean distance of 7934 metres) is never optimal as it would be for a noxious facility, even with maximal nuisances related to a recycling centre. The introduction of external costs in the objective function causes a remoteness of the optimal location of the service, but the loss of accessibility is limited (quite low nuisances are well accepted by the population).

10.3.2 Further Experiments

The location of several facilities

Several simulations were undertaken considering values of k greater than 1 (see Flahaut, 1997). If this is far from being realistic in terms of land-use planning at the time of the study, considering several polluting facilities enables us to see how far competing polluting facilities interact spatially. Figure 10.2 and Table 10.2 show the changes in average and maximal distance that users would have to travel if the number k of facilities increases. We see that $k = 3$ would be the 'best' number of facilities to locate.

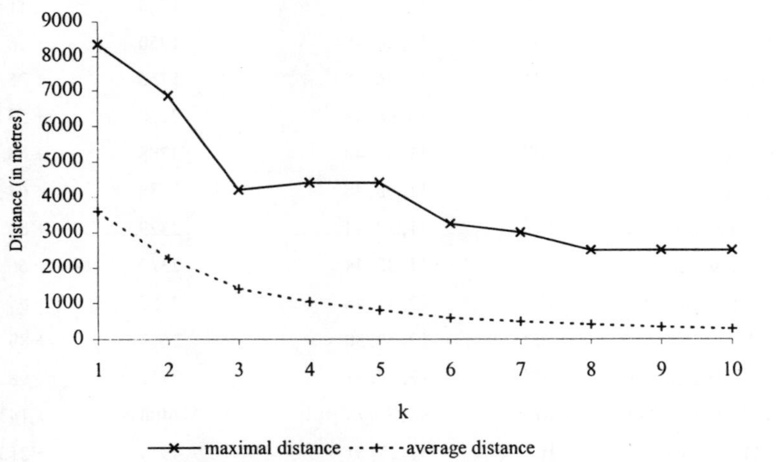

Source: Flahaut (1997)

Figure 10.2 Relationship between the number of facilities (k) and the average and maximal distances

Changes in the optimal locations with the increase of the negative externalities follow the same trend as with $k = 1$ (increase of the mean distance with an increase of the nuisances), but they take place earlier (that means with lower external costs) because of a better covering of the territory with three facilities instead of just one. The influence of the external costs on the optimal locations is more important than with $k = 1$ because the average distance between a supply and a demand node decreases when increasing the number of facilities. This has the effect of increasing the number of people located in the range of action of the nuisances and hence the sum of related external costs. External costs have at this moment more importance and their effect on optimal locations is consequently higher: the accessibility criterion is rapidly replaced by the search for less exposure to the nuisances.

Table 10.2 Main results for k = 3

WTP (in BEF)	Diffusion factor *C*	Optimal locations (index of nodes)	Average distance (single way, in metres)	(1)
-	-	15, 34, 45	1400	0
≥ 8422 & ≤ 20 746	10^{-4}	15, 34, 8	1432	+ 2
23 309	10^{-4}	15, 36, 8	1560	+ 11
≥ 24 751 & ≤ 42 293	10^{-4}	15, 36, 48	1750	+ 25
≥ 7598 & ≤ 8422	10^{-5}	15, 36, 48	1750	+ 25
47 740	10^{-4}	15, 38, 48	2150	+ 54
≥ 9887 & ≤ 20 746	10^{-5}	11, 38, 48	2288	+ 63
≥ 7598 & ≤ 8422	10^{-6}	11, 38, 48	2288	+ 63
≥ 58 313 & ≤ 79 780	10^{-4}	11, 38, 44	2329	+ 66
≥ 23 309 & ≤ 24 751	10^{-5}	11, 38, 44	2329	+ 66
≥ 9887 & ≤ 16 741	10^{-6}	12, 37, 48	2587	+ 85
19 304	10^{-6}	12, 37, 50	2653	+ 89
≥ 20 746 & ≤ 24 751	10^{-6}	12, 37, 51	2772	+ 98
≥ 26 273 & ≤ 79 780	10^{-5}	9, 37, 27	3406	+ 143
≥ 26 273 & ≤ 42 293	10^{-6}	37, 27, 51	4375	+ 213
≥ 47 740 & ≤ 79 780	10^{-6}	37, 38, 51	4990	+ 256

Note: (1) = Mean distance variation related to the optimal location for the *k*-median (spatial units nos 15, 34, 45), expressed in percentage

Source: Flahaut (1997)

The comparison of these results with those obtained on theoretical networks (Chapter 8) leads to similar conclusions: increasing the number of facilities for a given level of externalities leads to larger effects on location results than for a unique location, because these facilities are moved away from each other in order to avoid an addition of external costs generated by several undesirable facilities. This effect increases the remoteness of the facilities from the densely inhabited centres. With the highest external costs, two facilities out of three are concentrated in the north-western part of the studied area, that is, one of the less inhabited part of the 'commune'; if we further increase the nuisances (not realistic for a recycling centre), the three facilities are concentrated in this north-western part of the studied area (nodes 27, 37 and 38). This observed concentration is to be explained by the 'isolation effect', which means that the existence of a frontier does not allow us to consider population centres outside the limits of the studied area.

Another measure of demand

Until now, demand for a community recycling centre was considered as being best represented by the total population in each spatial unit. Is this really the case?

A restricted survey realised in 1996 in another Walloon township (Bureau Economique de la Province de Namur, 1996) is the only available source of data that gives us some indications about the kind of people who use recycling centres. The classification of people is based upon the size of the family, professional activity and age; Table 10.3 gives a summary of this classification. We used the proportions expressed in Table 10.3 and weighted the different socio-economic classes of the population. Hence, three different weights expressing demand were obtained. Each one leads to a set of simulations. In the case of one facility ($k = 1$), no difference is observed: the way of measuring demand does not modify the preceding conclusions about the effects of increasing externality costs in the location results. When locating three facilities ($k = 3$), only one spatial configuration slightly differs with the way of measuring demand: the configuration (sites 11, 38, 48) is replaced by (sites 12, 37, 48), with equivalent levels of nuisances. Hence, spending time in order to accurately estimate demand would not have been profitable, at least for this particular problem.

Table 10.3 Characteristics of the users of a recycling centre

Category		Users of a recycling centre (%)	Population Census (1991) (%)	Specialisation index
Household size	1	6	24	0.23
	2	33	29	1.15
	3	20	20	0.98
	4	25	16	1.52
	5	9	7	1.39
	≥ 6	7	3	2.18
Age	< 19	-	-	-
	19 to 29	11	20	0.59
	30 to 39	26	21	1.25
	40 to 49	35	17	2.02
	50 to 59	14	14	0.99
	≥ 60	14	26	0.52
Working activity	Employees	43	28	1.55
	Workers	12	12	1.03
	Indep./liberal profession	11	7	1.60
	Retired	16	28	0.59
	Not working	6	14	0.42
	Others	12	-	-

Source: Bureau Economique de la Province de Namur (1996) and Flahaut (1997)

An equal demand in each point is also considered ($w_i = 1, \forall\ i$). Results are here different from those observed with weighted demand especially when external costs are non-existent ($E_{ij} = 0$) or when they are very low: optimal locations are further away from the inhabitants. When external costs increase, results are quite similar. This situation can be explained as follows: when external costs are low in the case of a weighted demand, population centres have a large attraction that limits their action and this attraction disappears when using identical weights. For higher external costs, this attraction disappears because pollution is then too important to allow proximity of the facility to many people. These differences show the usefulness of using a weighted demand, and not an identical one, in the analysis. The tendency observed in both cases (weighted or identical demand) is nevertheless similar and consists in the optimal location of the facilities in the periphery of the studied area. This is normal with a uniform demand, but not necessarily with a weighted one: the spatial distribution of demand between spatial units should have an important role to play and it should therefore be interesting to study it.

An operational note

This analysis was undertaken independently from the concerned local authority. Meanwhile, node 13 has been selected by the authorities for locating the community recycling centre. There was by chance some land available. However, this choice corresponds to a solution where more attention is paid to accessibility than to pollution, because the mean distance is not much larger than the one computed for node 15, which is optimal for the *k*-median (4187 m / 3600 m). A positive aspect is nevertheless that node 13 is not densely inhabited despite a quite central position (near the community centre of Rhisnes). Local authorities have taken additional criteria into account for choosing this location: the proximity of an important national road and the availability of land that could be used to locate such an infrastructure. The choice of a location close to the population such as node 13 can even better be justified when more people use the service: the more people use it, the less they are affected by its semi-obnoxious character because they attribute more importance to its usefulness. Hence, it is necessary to take measures that encourage people to use the community recycling centre, in order to justify a location quite close to population centres.

The frequency of use of the service by the population is the most important factor influencing the evaluation of emitted nuisances (no diffusion is here considered). Indeed, transport costs and external costs of nuisances are summed in the objective function (see Section 10.2.2); for a given period of time, transport costs depend upon the number of visits to the recycling centre. The more frequent are the visits to the service, the less the external costs are important in comparison to transport costs: the more the service is used by the population, the fewer inhabitants consider this service as undesirable. This variation of the perceived nuisance is related to the semi-obnoxious nature of the recycling centre, for which pollution is relatively well accepted by the people.

10.4 CONCLUDING REMARKS

This small example confirms the results obtained in Chapter 8 and shows how far the optimal location of a semi-obnoxious facility (a community recycling centre) is sensitive to the importance of the externalities. Within the limits of the Belgian land-use planning case-study, we confirm that only medium- and high- externality measures significantly change the optimal location results. Optimal locations are a compromise between transportation and pollution costs; transportation costs no longer structure the optimal results when important negative externalities are important. In land-use planning circumstances, where data about perceived and real pollution are scarce or of poor quality, it is useful to know this: small errors in the measure of externalities do not influence the location-allocation results. At the same time, this chapter illustrates the operationality of the adopted bi-objective model for solving local planning problems of polluting facilities: the model is fully operational and easy to use. It is important to underline that the results presented here depend upon numerous assumptions, especially those concerning the model and the estimation of the external costs.

11. Conclusion

It is our belief that obtaining general results regarding the impact of the transport network on the location of human activities is quite ambitious and highly problematic because of the combinatorial nature of the problem, even in the case of toy-networks. However, we believe that the kind of experimental approach followed in these few pages leads to some interesting results because it relies more on what we know from geography and spatial microeconomics. The objective of this last chapter is twofold: the first is to highlight the major contributions of the preceding chapters in a succinct manner and to suggest policy recommendations (Section 11.1), the second is to suggest some avenues for future research that this book has brought to light (Section 11.2).

11.1 MAIN RESULTS AND POLICY RECOMMENDATIONS

In this research, we have considered the impact of the transportation spatial structure on the optimal locations of human activities. More precisely, we have considered the spatial characteristics of the transportation network and analysed how far they influence the results of some optimal location-allocation models (the *k*-median and the UFLP). With the exception of Chapters 9 and 10 that rely on real-world land-use planning circumstances, results are based on simulations performed on theoretical networks (toy-networks designed on squared or hexagonal lattices of points).

It is worth recalling that the initial methodological objective was to correlate indices depicting the shape of the network to quantitative measures associated with the output of the location-allocation models. Chapter 2 has shown that at the beginning of the twenty-first century, it is still impossible to associate to a given transport network characterised by more than two nodes, a synthetic index that reflects unequivocally its shape and size (its spatial morphology). Given the complexity of the problem, it can even be impossible to quantify. Hence, the relationship between the transportation network and location-allocation results has not been quantified, nor statistically tested, but just described and illustrated by means of tables, figures and maps. The numerical results mentioned in Chapters 3 to 10 depend on the various assumptions about the parameters, the many simplifications applied when measuring the location inputs and on the ways the scenarios have been formulated. Most results however point in the same direction. Of course, this does not preclude the possibility of getting opposite results with completely different data sets (for example discontinuous space, congestion-sensitive networks, location-specific production costs, actual distances, larger lattices, and so on). We, however, believe that the main patterns emerging from the experiments are fairly robust and contribute to a better operationality of the location-allocation models. The main results can be summarised as follows.

(1) Independently of the connectivity of the network, there is a strong relationship between the shape of the network and the optimal spatial organisation of activities, whatever the studied output of the model (locations, service areas, and so on), whatever the type of underlying lattice of points (square, hexagon), or whatever the location technique (*k*-median, UFLP). More precisely, it has been shown that radial networks lead to more concentrated patterns of production, the centre exerting strong attractive forces. In contrast, a grid network leads to a more dispersed pattern with several locales accommodating facilities. Though such results belong to the folk wisdom of human geography, the many experiments give them more robustness (Chapter 3). The results for the radial network were particularly strong, because the model did not include any capacity constraint and did not consider congestion. The general role of the morphology of the transportation is further illustrated in Chapter 9 for a Third World planning problem: a location-allocation model for health care facilities is applied separately for the rainy season and for the dry season. Each season assuming a different operational transportation network, the comparison of the modelling results shows that the suggested number of optimal locations is much greater in the rainy season than in the dry season and hence, that the shape of the network it is important to the operational results. Land-use planners should be aware of this when planning facilities, as well as transportation networks.

(2) Interesting enough, the first conclusion still holds in integrated economies (Chapter 4). Hence, to the extent that the construction of transportation infrastructures can be regarded as irrevocable, networks are hostages of the past, and continue to influence the decisions made in the new economic and/or political environment. This legacy is heavy: within the limits of the regional economy, a given spatial morphology of a network has about the same impact on the spatial organisation of activities in the case of autarky as in an integrated economy. Hence, even though the level of activity is likely to be affected by the integration of two economies, it is less obvious that the spatial organisation of human activities within each regional

economy is to be drastically modified: the geographical organisation of economic and political activities might well remain approximately the same within each original territory. Accordingly, one must be careful before drawing clear-cut implications about the impact of economic integration on the spatial distribution of human activities.

(3) An interesting result that has been put forward in Chapters 3 and 4 is the role of the adjunction of a peripheral ring road on a given transportation network. Adding a peripheral link to a radial network modifies the shape of the transport network, but also generates important changes in the location results. The addition of a peripheral link reveals how the attractiveness of the centre may be weakened or reinforced, depending on the location of the ring road. The construction of a peripheral road around the centre of a radial network can boost a major deconcentration of human activities towards the nodes situated at the crossings between the radial roads and the ring road. If such a policy is to be implemented, the choice of the ring radius turns out to be a critical policy parameter: a too-small radius leads to a circumradial network that does not depart sufficiently from the original radial configuration to generate dispersion, while a very large radius yields nodes with hinterlands too small to become attractive. Clearly, even when the radius of the peripheral road is well chosen, it would be illusive to expect the outward pull discussed above to generate its effects in the short or medium run. The possible impact of such a policy on locations will become visible only in the (very) long run. These are only hypotheses, as no time component has been introduced in the models.

The findings about the ring road has however important implications for further empirical planning work: when location is of concern, the planner should be aware of the importance of the location and size of a peripheral road, and more generally to how the addition of one or several new links modifies the structure of the transportation network.

(4) Merging two economies in a common market implies consideration of the way they are merged: the number of gates between the two economies, and the location of those gates and/or the 'length' associated with them, define the efficiency of the border. Chapter 5 shows that the influence of the characteristics of the border in an integrated economy is less interesting than expected: location results have little sensitivity to changes in the permeability of the border expressed in the number and location of the gates. The 'length' of the gate has the most important consequences for the location-allocation results; this puts forward the role plaid by tariffs and/or congestion at some border points in the previous and the present European regional contexts.

(5) Chapter 5 contains interesting results about the effect of a high-speed connection between two regional centres in an integrated economy: the building of a high-speed connection between two regional centres gives rise

to some distinctive geographical features. Those can be summarised as follows: firstly, the high-speed connection does affect the urban hierarchy, but not necessarily in a dramatic way. Secondly, the high-speed connection reduces the lure of the gateway city between the constituent economies. Thirdly, the high-speed connection fosters some regional imbalance since one centre serves a growing market segment in the other region. Ultimately, one regional centre can become the only production centre for values of the fixed production costs which are not that high (compared to values of fixed costs obtained in the absence of the high-speed connection).

The simulations have, however, revealed some structural instability of the locational pattern as the value of the fixed production costs rises relative to that of the transport costs. This seems to confirm the scepticism of some transportation analysts in that it appears to be hard to predict what the final impact of a real-world infrastructure will be.

(6) Implementing a GIS for applied location-allocation techniques (or even for any other geographical model) often requires the construction of a large distance matrix. Distances can be directly measured on a map, but can also be estimated. In Chapter 6 we simply show how robust the solutions of a location-allocation model are to changes in the parameters of one particular distance-predicting function (the l_p-distance) and consequently with what accuracy these parameters should be estimated in real-world applications. The simulations undertaken in this research lead to the following conclusions: changes in the l_p-distance parameters influence the absolute values of the location parameters (such as the maximum or the average distance), but have little effect on the location results (the choice of the location sites). Practical implications arising from these simulations are twofold. Firstly, small errors in the empirical estimation of distances by the l_p-distance-predicting function very slightly affect the location-allocation results. This is of prime importance when considering the amount of work involved in estimating the real-world distances by l_p-distances: it saves data collecting and computing time whenever such applications are considered. In other words, the substantial volume of extra work involved in determining the exact parameters for the l_p-distance estimation function is not really worth it in location-allocation modelling as it is gives only a slender chance of obtaining more accurate or different results. Of course, this conclusion does not hold if the value of the l_p parameters differs for each link of the network. Secondly, in accordance with the preceding conclusions, Chapter 6 seems to confirm that the coordinates of the points, and hence the shape of the transportation network, have more influence on the k-median results than the way the distance is measured.

(7) Chapter 7 deals with the integration of spatial pricing strategies in location-allocation models, with the financing of the transportation costs.

Our purpose is to try to disentangle the relative influences of the shape of the transportation network, of the economies of scales in production, and of the geographical pricing system on the location pattern of producing facilities of a firm, on its spatial schedule of prices, and thus on the characteristics of its market area. Within the limits of the conducted sensitivity analyses (small-sized toy-networks, some shapes of transportation networks, squared lattice of points, and so on), the principles that seem to emerge can be summarised as follows: the choice of a price policy has a limited impact on firms' location, compared to the impact of the transport network configuration. This suggests that shaping the transport network proves to be a rather powerful instrument in spatial planning, especially in countries with limited infrastructures like the less developed countries.

(8) In Chapter 8, a model that minimises both transportation costs and pollution costs is considered. Within the limits of the experiments, the performed simulations show that all locations changes are generated by changes in the balance between transportation costs and polluting costs: slight pollution does not modify the solutions of the location-allocation model; high extension or intensity of the pollution generates 'competition' between facilities and modifies the reference locations. Facilities are located further apart from each other and hence further from the consumers. Very high values of the intensity and/or the extension tend to push the optimal locations to the border of the studied environment. In other words, when the pollution is slight, accessibility of the supply sites remains the most important location criterion. When pollution gets more important, facilities tend to be located as far as possible from each other, tending to another spatial optimum. Applied work should be aware of this when finding the best spatial solution: pollution is not always a source of optimality changes. For our purposes, we see that very large externalities annihilate transportation costs and lead to locations on the borders of the studied area. Hence, when transportation costs are not the unique criterion that is optimised, the structuring role of the transportation network is less that the one observed and discussed in the preceding chapters.

These last results were illustrated with a real-world application in Chapter 10: the optimal location of a community recycling centre is studied in a Belgian residential township (La Bruyère, Wallonia). The only pollution cost introduced is the noise. It is shown that accessibility is the most important criterion for determining the location; only very high levels of pollution push the recycling centre far away from the population centres. Note also that, as pollution is often quite difficult to estimate and measure, a small measurement error should not have important operational consequences as far as locations are concerned.

Given the limits of this research (some toy-networks, some location models, many simplifying assumptions, some values of the parameters), the conclusions derived from the models are suggestive rather than conclusive, and they need to be interpreted with caution in a policy context. However, the main results of this research suggest that major transport facilities are often considered as a favouring item in town and country planning. Yet, the decisions concerning large projects are most of the time taken on the basis of financial and sometimes political criteria (Klein, 1998). Arguments in terms of town and country planning are very rarely sufficiently reasonable to convince those scientists who usually work on the transport-space theme. At the same time, the reappraisal of the structuring effect concept (Offner, 1993) does not seem to give rise to any reaction or even any debate among the same scientists. The results suggested here should help the decision-makers to reappraise the spatial structuring role of the transportation infrastructure, especially in developing regions or countries.

Given the large body of work already published on the relationship between the general provision in transportation infrastructure and land-use or urban or regional development (Chapter 1), we are far from asserting that every aspect in the domain has been covered and solved throughout this book. We just hope that the outcome of this research functions as a small stepping stone in the elaboration of a more general theory in economic geography, and (2) helps to better the understanding of the sensitivity of the results of the location-allocation models to the measure of their inputs, and more particularly to the transportation costs and structure.

11.2 SUGGESTIONS FOR FURTHER RESEARCH

At this stage, we are aware that the research scope has been rather narrow. The analysis is incomplete in several respects: the simulations were restricted to two location-allocation models, to toy-networks, to two shapes of the lattices (square and hexagon), to small-sized lattices, to a limited number of networks (grid and radial), to common markets with only two regional economies of equal size, to transportation costs limited to physical distances, to experimental designs with no competition between modes of travel or between goods or services, to the chosen amplitude of variation of the different parameters, and so on. These choices were made among an infinity of alternatives. It is not certain that modifying those choices will lead to the same conclusions as those suggested here. In this sense, the number of possible extensions that are worth studying is enormous. Let us here cite some of the most relevant extensions, classified in three broad categories.

Technical Extensions

Even if the location-allocation techniques have largely been improved over the past twenty years (Chapter 1), there is still a need for further extensions of the models and of the solution algorithms. At several stages in our research project, we were limited by the size of the toy-networks (see for example Chapter 7), by the lack in sufficient operational constraints (Chapters 8 and 10), and by the existence of quite disturbing alternative optima (Chapter 6). These resolution problems should be overcome in the future. Moreover, the operational models used in this book are mainly of the equilibrium type, sometimes with possibilities for treating the development over time in a recursive manner. The further development of dynamic facility sizing and location is another (technical) area of research that deserves increased attention, especially as dynamic models may provide better planning tools in many instances.

A second interesting field for further 'technical research', relies on GIS. There is indeed an increasing availability of georeferenced data and geographical information systems technologies; these should favour modelling contexts in land-use planning. Therefore, the integration of operational location-allocation models in GIS is a viable research programme in its own right. Some attempts already exist (for example Birkin et al., 1996; Arentze, Borgers and Timmermans, 1996; Miller, 1996; Hamacher et al., 1999), but they are still limited to simple formulations of the models, with little operational choices. The simulations performed here illustrate the interest in such a research topic.

Statistical Extensions

As already mentioned in the preceding chapters, although a strong need remains for continued development and enhancement of sensitivity analyses such as those conducted in this research, there is a need to develop better methodologies for comparing location-allocation results, for evaluating them scientifically and for testing them statistically. Up to now and as far as we know, no statistical test has been used and developed in the literature. Hence, the simulations performed in the preceding chapters were limited to the descriptive comparisons of measurable outputs of the models.

Economic Geography Extensions

An important aspect of this book is that it is based on the assumption that total demand is given (population, employment); this means that it can only deal with the spatial distribution of activities, not with the total level of activities. Generative effects on employment, investments, demographic changes, and so on are left out of consideration. The reason is of course that for this purpose one needs insight into the locational behaviour of firms or households in an inter-regional orientation. This would entail the development of linkages between transport land-use models and inter-regional trade, as well as location models, that is to say a topic that was far beyond the scope of this analysis, but that can be considered as a promising direction for future development

The spatial distribution of demand is another topic that undoubtedly deserves more attention, especially in those areas large enough to be to some extent self-supporting. In the chapters developed in this book, the spatial distribution of demand was very simplistic: it was considered as uniformly distributed, with the exception of Chapters 4 and 7. In Chapter 7, demand was endogenously defined, and in Chapter 4 a uniform distribution of demand was compared to a random and to a 'pyramidal' distribution (centre-periphery). More case-studies should be analysed: what happens, for instance, if demand is concentrated in the peripheral areas of a radial or circumradial network? What happens when the amplitude of demand varies much more spatially? As previously mentioned, what happens when the relocation of the households or firms is made possible and introduced in the model? For example, the question whether an improvement in transport infrastructure in an area makes this area more competitive so that it attracts more investments or households than it would otherwise, has here not been considered.

Hence, taking into account demand mechanisms and relocation would consideration of the link between this topic and the evolution of the emergence of the urban systems. Considering a dynamic process would enable the linkage of transportation to urban development and to the economies of agglomeration, but also to the diseconomies such as commuting and hence congestion. Transport infrastructure is indeed one of the instruments of land-use planning, but there should be an important difference between the availability of the infrastructure and its use, especially as far as car use is concerned.

Most operational results were here interpreted at the regional level. It should be interesting to change the scale of analysis to smaller units such as individual urban entities, to conurbations, or even to intra-industrial spatial organisations. Understanding the effect of the structure of the transportation network on the intra-urban structure would, however, force us to include congestion in the model and hence (may be) modulate some of the results presented here.

Last but not least, it should be important to buttress the theoretical approach by more complex empirical work, and to situate this research in a broader context of research issues regarding societal and economic changes, and transport (for example Giuliano and Gillepsie, 1997): this book was restricted to the shape of the transportation network without any causal link with the forces driving this shape. It would be interesting to see how far the current major spatial and societal trends are bound to changes in the location of activities and to the demand for transport, growth of the elderly population, foreign immigration, decline of the nuclear family, growing income disparities and their consequent spatial residential choices, impacts of information technology on lifestyle, teleworking, increase in flexible forms of employment, changing linkages between and among firms, the renewal of the city centres, and so on). These not only influence transport, but also the distribution of the activities. In most of this book, we let the transportation network vary for an exogenously given spatial distribution of demand; this is quite artificial. Should not we further consider the variation of demand as well? How far do the agglomeration forces interact with transportation? What are the driving forces in the spatial organisation: the locations of the human activities, or the characteristics of the transportation network, or both? How far does urban development influence transportation infrastructure and vice versa? Can those effects be isolated at each scale of analysis, or are they interrelated in such a way that this is just a chicken and egg problem?

In this research, the infrastructure investments were artificially isolated from other relevant land-use developments. In other words, it is now time to relax some of the assumptions, to let other sources of variation interact in order to better understand the role of transport on the optimal location of activities, and vice versa. This however requests a strong collaboration between the various disciplines concerned (economics, computer science, geography, mathematical programming, statistics) in theoretical as well as in empirical terms, in modelling as well as in non-modelling terms. Multidisciplinarity and transversality are the two leading trends for further constructive research in this fascinating, but quite complex research topic.

References

Abdel-Rahman, M. (1996), 'When do cities specialize in production?', *Regional Science and Urban Economics,* **26**, 1-22.

Airey, T. (1991), 'The influence of road construction on the health care behaviour of rural households in the Meru District of Kenya', *Transport Reviews,* **11,** 273-90.

Alao, N., Dacey, M., Davies, O., Denike, K., Huff, J., Parr, J. and Webber, M. (1977), *Christaller Central Place Structure: An Introductory Statement*, Evanston, Northwestern.

Algoed S. (1996), *Analyse de l'impact des politiques spatiales de prix et de la morphologie des réseaux sur la localisation d'une entreprise en situation de monopole*, UCL, Louvain-la-Neuve, Mémoire en sciences économiques.

Allen, G. (1981), 'A developmental perspective on the effect of subdividing macrospatial experience', *Journal of Experimental Psychology,* **7**, 120-32.

Allen, W., Liu, D. and Singer, S. (1993), 'Accessibility measures of US metropolitan areas' *Transportation Research,* **27B**(6), 439-49.

Alonso, W. (1969), 'Urban and regional imbalances in economic developments', *Economic Development and Cultural Changes,* **17,** 1-14.

Amoud, I. (1999), *Localisation des dispensaires ruraux dans l'arrondissement de Filingué (Niger)*, mémoire inédit de licence UCL en Architecture Urbaine, Louvain-la-Neuve, 127 p.

Anas, A. (1987), *Modelling in Urban and Regional Economics*, New York: Harwood.

Anas, A., Arnott, R. and Small, K. (1998), 'Urban spatial structure', *Journal of Economic Literature,* **36,** 1426-64.

Anderson, W., Kanaroglou, P. and Miller, E. (1996), 'Urban form, energy and the environment: a review of issues, evidence and policy', *Urban Studies,* **33***(1),* 7-35.

Anselin, L. (1986), 'Some further notes on spatial models and regional science', *Journal of Regional Science,* **26**(4), 799-802.

Anselin, L. (1998), 'Exploratory spatial data analysis in a geocomputational environment', in Longley, P., Brooks, S., McDonnell, R. and Macmillan, B. (eds), *Geocomputation: A Primer*, Chichester: John Wiley, pp. 77-94.

Aoyagi, M. and Okabe, A. (1993), 'Spatial competition of firms in a bounded two-dimensional market', *Regional Science and Urban Economics*, **23**, 259-89.

Arbia, G. (1989), *Spatial Data Configuration in Statistical Analysis of Regional Economic and Related Problems*, Dordrecht: Kluwer Academic Publishers, 257 p.

Arentze, T., Borgers, A. and Timmermans, J. (1996), 'Design of a view-based DSS for location planning', *International Journal of Geographical Information Systems*, **10**, 219-36.

Armstrong, M. (2000), 'Geography and computational science', *Annals of the Association of American geographers*, **90**(1), 146-56.

Arnold, P. (1994), *Effets de l'organisation spatiale du réseau de transport sur les résultats du modèle de la p-médiane*, mémoire inédit de licence UCL en Architecture Urbaine, Louvain-la-Neuve, juin 1994, 58 p.

Arnold, P. and Peeters, D. (1995), 'Sensibilité du modèle de la *p*-médiane au comportement spatial de l'usager', *L'Espace Géographique*, **24**(1), 25-35.

Arthur, W. (1990), 'Silicon Valley locational clusters: when do increasing returns imply monopoly ?', *Mathematical Social Sciences*, **19**, 235-51.

Aschauer, D. (1989), 'Is public expenditure productive?', *Journal of Monetary Economics*, **23**(2), 177-200.

Attneave, F. and Arnoult, M. (1966), 'The quantitative study of shape and pattern perception', in Uhr, L. (ed.), *Pattern Recognition: Theory, Experiment, Computer Simulations, and Dynamic Models of Form Perception and Discovery*, New York: John Wiley and Sons.

Auphan, E. (1992), 'Les gares TGV régionales: un exemple de contre aménagement du territoire', *Hommes et Terres du Nord*, **92**(1), 14-20.

Ayeni, B., Rushton, G. and McNulty, M. (1987), 'Improving the geographical accessibility of health care in rural areas: a Nigerian case study', *Journal of Social Sciences and Medicine*, **25**, 1083-94.

Bailey, T. and Gatrell, A. (1995), *Interactive Spatial Data Analysis*, Harlow: Longman.

Bailly, A. (1985), 'Distances et espaces: vingt ans de géographie des représentations', *L'Espace Géographique*, **85**(3), 197-205.

Bairoch, P. (1985), *De Jéricho à Mexico. Villes et économie dans l'histoire*, Paris : Gallimard. English translation: *Cities and Economic Development: From the Dawn of History to the Present*, Chicago: University of Chicago Press.

Barwell, J., Gould, B., Clarke, C., Mather, S., Prothero, M., Siddle, D., Smith, C. and Thomas-Hope, E. (1996), *A Geography of the Third World*, London and New York: Routledge, 344 p.

Batty, M. (1989), 'Urban modelling and planning: reflections, retrodictions and prescriptions', in Macmillan, B. (ed.), *Remodelling Geography*, Oxford: Blackwell, 147-69.

Batty, M. (1994), 'A chronicle of scientific planning: the Anglo-American modeling experience', *Journal of the American Planning Association*, **60**, 7-16.

Batty, M. and Longley, P. (1994), *Fractal Cities, a Geometry of Form and Function*, Academic Press.

Baumol, W. (1980), 'Contestable markets: an uprising in the theory of industry structure', *American Economic Review*, **70**, 1-15.

Beckmann, M. (1976), 'Spatial price policies revisited', *Bell Journal of Economics*, **7**, 619-29.

Beckmann, M. (1985), 'Spatial price policy and the demand for transportation', *Journal of Regional Science*, **25**, 367-71.

Beckmann, M. and Ingene, C.A. (1986), 'The profit equivalence of mill and uniform pricing policies', *Regional Science and Urban Economics*, **6**, 327-29.

Beckmann, M. and Thisse, J.-F. (1986), 'The location of production activities', in Nijkamp, P. (ed.), *Handbook of Regional Economics*, Amsterdam: North-Holland, pp. 21-95.

Begg, R. (1992), 'Social and spatial context in rural development strategies', *International Regional Science Review*, **15**, 1-11.

Beguin, H. and Thomas, I. (1997), 'Morphologie du réseau de communication et localisation optimales d'activités: quelle mesure pour exprimer la forme d'un réseau?', *Cybergéo* (Revue Européenne de Géographie sur Internet), paper 26.

Beguin, H. and Thisse, J.-F. (1979), 'An axiomatic approach to geographical space', *Geographical Analysis*, **11**, 325-41.

Beguin, H., Thomas, I. and Vandenbussche, D. (1992), 'Weight variations within a set of demand points, and location-allocation issues: a case study of public libraries', *Environment and Planning,* **24A**, 1769-79.

Ben-Akiva, M. and Lerman, S. (1985), *Discrete Choice Analysis: Theory and Application to Travel Demand*, Cambridge, MA: MIT Press.

Benguigui, L. (1995), 'A fractal analysis of the public transportation system of Paris', *Environment and Planning*, **27A**, 1147-61

Benguigui, L. and Daoud, M. (1991), 'Is the suburban railway system a fractal?' *Geographical Analysis*, **23**, 362-68.

Bennett, V., Eaton, D. and Church, R. (1987), 'Selecting sites for rural health workers', *Journal of Social Sciences and Medicine*, **16**, 1072-94.

Berechman, J. (1994), 'Urban and regional impacts of transportation Investment: a critical assessment and proposed methodology', *Transportation Research*, **28A** (4), 351-62.

Berechman, J., Kohno, H., Button, K. and Nijkamp, P. (eds) (1996), *Transport and Land Use*, Modern Classics in Regional Science: 2, Cheltenham and Brookfield: Edward Elgar Publishing.

Berens, W. and Körling, F. (1985), 'Estimating road distances by mathematical functions', *European Journal of Operational Research*, **21**, 54-6.

Berens, W. and Körling, F. (1988), 'On estimating road distances by mathematical functions – a rejoinder', *European Journal of Operational Research*, **36**, 254-5.

Berge, C. (1963), *Théorie des graphes et ses applications*, Paris: Dunod.

Berge, C. (1973), *Graphs and Hypergraphs*, Amsterdam: North Holland. French version (1973), *Graphes et hypergraphes*, Paris: Dunod.

Berman, O., Ingco, D. and Odoni, A. (1992), 'Improving the location of minisum facilities through network modification', *Annals of Operations Research*, **40**, 1-16.

Berman, O., Ingco, D. and Odoni, A. (1994), 'Improving the location of minimax facilities through network modification', *Networks*, **24** (1), 31-41.

Berman, O., Larson, R. and Chiu, S. (1985), 'Optimal server location on a network operating as an M/G/1 queue', *Operations Research*, **33**, 746-70.

Berry, B. and Marble, D. (1968), *Spatial Analysis: A Reader in Statistical Geography*, New Jersey: Prentice-Hall.

Bertuglia, C., Clarke, G. and Wilson, A. (1994), 'Models and performance indicators in urban planning: the changing policy context', in Bertuglia, C., Clarke, G. and Wilson, A. (eds), *Planning and Analysis in Health Care Systems*, London: Pion, pp. 90-118.

Birkin, M., Clarke, G., Clarke, M. and Wilson, A. (1996), *Intelligent G.I.S. Location Decisions and Strategic Planning*, Cambridge: GeoInformation International, 292 p.

Blum, U., Gercek, H. and Viegas, J. (1992), 'High-speed railway and the European peripheries: opportunities and challenges', *Transportation Research*, **26A**, 211-21.

Boffey, B. and Karkazis, J. (1993), 'Models and methods for location and routing decisions relating to hazardous materials', *Studies in Locational Analysis*, **5**, 149-66.

Boffey, B. and Karkazis, J. (1995), 'Location, routing and the environment', in Drezner, Z. (ed.), *Facility Location: A Survey of Applications and Methods*, New York: Springer, pp. 453-66.

Bollobas, B. (1990), *Graph Theory: An Introductory Course*, Berlin: Springer.

Bonnafous, A. (1989), 'The regional impact of the TGV', *Transportation*, **14**(2), 127-37.

Bonnafous, A. (1994), 'Réseaux de transport', in Auray J.-P., Bailly A., Derycke, P.-H. and Huriot J.M. (eds), *Encyclopédie d'Economie Spatiale*, Paris: Economica, Bibliothèque de Science Régionale, pp. 325-32.

Boots, B. (1984), 'Evaluating principal eigenvalues as measures of network structure', *Geographical Analysis*, **16**(3), 270-75.

Boots, B. (1985), 'Size effects in the spatial patterning of nonprincipal vectors of planar networks', *Geographical Analysis*, **17**(1), 74-81.

Boots B. and Royle, G. (1991), 'A conjecture on the maximum value of the principal eigenvalue of a planar graph', *Geographical Analysis*, **23**(3), 276-82.

Braeutigam, R., Daughety, A. and Turnquist, M. (1984), 'A firm specific analysis of economies of density in the US railroad industry', *Journal of Industrial Economics*, **33**, 3-20.

Brandeau, M. and Chiu, S. (1989), 'An overview of representative problems in location research', *Management Science*, **35**, 645-73.

Brimberg, J. and Love, R. (1992), 'A new distance function for modelling travel distances in a transportation network', *Transportation Science*, **26**(2), 127-37.

Brimberg, J. and Love, R. (1993a), 'Directional bias of the l_p norm', *European Journal of Operational Research*, **67**, 287-94.

Brimberg, J. and Love, R. (1993b), 'General considerations on the use of the weighted l_p norm as an empirical distance measure', *Transportation Science*, **27**(4), 341-49.

Brimberg, J. and Wesolovsky, G. (1992), 'Probabilistic *lp* distances in location models', *Annals of Operations Research*, **40**, 67-75.

Bruinsma, F. and Rietveld, P. (1993), 'Urban agglomerations in European infrastructure networks', *Urban Studies*, **30**(6), 919-34.

Brulard, T. and Van der Haegen, H. (1972), 'Small area statistics and their use for social geographical and planological research', *Acta Geographica Lovaniensia*, **10**, 448 p.

Buchanan, J. and Stubblebine, W. (1962), 'Externality', *Economica*, **29**, 371-84.

Bureau Economique de la Province de Namur (BEPN), Département Environnement, Gestion et Valorisation des déchets (1996), Groupe d'étude sur la fréquentation des parcs à conteneurs: 'Comptes rendus' and 'Synthèse et conclusions'.

Busacker, R. and Saaty, T. (1965), *Finite Graphs and Networks: An Introduction with Applications*, New York: McGraw-Hill.

Button, K. (1993), *Transport Economics*, Aldershot and Brookfield, Edward Elgar Publishing.

Casillas, P. (1987), 'The aggregation problem in the p-median model', in Ghosh, A. and Rushton, G. (eds), *Spatial Analysis and Location-Allocation Models*, New York: Van Nostrand, pp. 327-44.

Casti, J. (1995), 'The Theory of Networks', in Batten, D., Casti, J. and Thord, R. (eds), *Networks in Action: Communication, Economics and Human Knowledge*, Berlin: Springer-Verlag, pp. 1-24.

Catin, M., (1985), *Effets externes, marché et systèmes de décision collective*, Paris: Cujas.

Catin, M. (1994), 'Externalités', in Auray, J.-P., Bailly, A., Derycke, P.-H. and Huriot, J.M. (eds), *Encyclopédie d'Economie Spatiale*, Paris: Economica, Bibliothèque de Science Régionale, pp. 99–103.

Chapman, K. and Walker, D. (1987), *Industrial Location*, Oxford: Basil Blackwell.

Chhajed, D., Francis, R. and Lowe, T. (1993), 'Contributions of operations research to location analysis', *Location Science*, **1**, 263-87.

Chorley, R. (1972), 'Spatial analysis in geomorphology', in Chorley, R. (ed.), *Spatial Analysis in Geomorphology*, London: Methuen, pp. 3-16.

Christaller, W. (1933), *Die Zentralen Orte in Süddeutschland*, Jena: Gustav Fischer Verlag. English translation (1966), *The Central Places of Southern Germany*, Englewood Cliffs (N, J): Prentice Hall.

Church, R. and Garfinkel, R. (1978), 'Locating an obnoxious facility on a network', *Transportation Science*, **12**, 107-18.

Clark, G., Feldman, M. and Gertler, M. (eds) (2000), *The Oxford Handbook of Economic Geography*, New York: Oxford University Press.

Clarke, C., Dickenson, J. and Gould, B. (1996), *A Geography of the Third World*, London: Routledge, 344 p.

Cliff, A. and Ord, J. (1981), *Spatial Processes*, London: Pion.

Cliff, A., Haggett, P. and Ord, K. (1979), 'Graph theory and geography', in Wilson, R. and Beineke, L. (eds), *Applications of Graph Theory*, London: Academic Press.

Cliff, A., Haggett, P., Ord, J., Bassett, K. and Davies, R. (1975), *Elements of Spatial Structure: A Quantitative Approach*, Cambridge: Cambridge University Press.

Cornuejols, G., Nemhauser, G. and Wolsey, L. (1990), 'The uncapacited facility location problem', in Francis, R.L. and Mirchandrani, P.B. (eds), *Discrete Location Theory*, New York: J. Wiley, p. 119-71.

Couclelis, P. (1998), 'Geocomputation in context', in Longley, P., Brooks, S., McDonnell, R. and Macmillan, B. (1998), *Geocomputation: A Primer*, Chichester: John Wiley, pp. 17-29.

Cressie, N. (1993), *Statistics for Spatial Data*, New York: John Wiley.

Cronon, W. (1991), *Nature's Metropolis: Chicago and the Great West*, New York: Norton.

Current, J. and Ratick, S. (1995), 'A model to assess risk, equity and efficiency in facility location and transportation of hazardous materials', *Location Science*, **3**(3), 187-202.

Current, J. and Shilling, D. (1987), 'Elimination of source A and B errors in *p*-Median location problems', *Geographical Analysis*, **19**, 95-110.

Dasarathy, B. and White, L. (1980), 'A maximin location problem', *Operations Research*, **28**, 1385-401.

Daskin, M. (1995), *Network and Discrete Location, Models, Algorithms and Applications*, New York: Wiley Interscience.

Daskin, M., Haghani, A., Khanal, M. and Malandraki, C. (1989), 'Aggregation effects in maximum covering models', *Annals of Operations Research*, **18**, 115-40.

Daskin, M., Hurter, A. and Van Buer, M. (1993), 'Toward an integrated model of facility location and transportation network design', *Working Paper, Transportation Center, Northwestern University*, 30 p.

de la Fuente, A. and Vives, X. (1995), 'Infrastructure and education as instruments of regional policy: evidence from Spain', *Economic Policy*, **20**, 13-51.

Dell'Amico, M., Maffioli, F. and Martello, S. (eds) (1997), *Annotated Bibliographies in Combinatorial Optimization*, Chichester: Wiley, 495 p.

Demange, G. and Henriet, D. (1991), 'Sustainable oligopolies', *Journal of Economic Theory*, **54**, 417-28.

Densham, P. and Rushton, G. (1992a), 'Strategies for solving large location-allocation problems by heuristic methods', *Environment and Planning*, **24A**, 289-304.

Densham, P. and Rushton, G. (1992b), 'A more efficient heuristic for solving large *p*-median problems', *Papers in Regional Science*, **71**, 307-29.

Dickenson, H. and Rushton, G. (1979), 'Spatial efficiency of service locations and the regional development process', *Papers of the Regional Science Association*, **42**, 83-97.

Dijkstra, E. (1959), 'A note on two problems in connection with graphs', *Numerische Mathematik*, **1**, 269-71.

Dijst, M. and Vidakovic, V. (1997), 'Individual action space in the city', in Ettema, D. and Timmermans, H. (eds), *Activity-based Approaches to Travel Analysis*, Oxford: Pergamon, pp. 117-34.

Downs, R. and Stea, D. (1977), *Maps in Mind: Reflections on Cognitive Mapping*, New York: Harper and Row.

Drezner, Z. (ed.) (1995), *Facility Location: A Survey of Applications and Methods*, Heidelberg: Springer Verlag, 571 p.

Drezner, Z. and Wesolowsky, G. (1983), 'Location of an obnoxious facility with rectangular distances', *Journal of Regional Science*, **23**, 241-48.

Drezner, Z. and Wesolowsky, G. (1985), 'Location of multiple obnoxious facilities', *Transportation Science*, **19**, 193-202.

Drezner, Z. and Wesolowsky, G. (1990), 'The Weber problem on the plane with some negative weights', *INFOR*, **29**, 87-99.

Drezner, Z. and Wesolowsky, G. (1996), 'Obnoxious facility location in the interior of a planar network', *Journal of Regional Science*, **35**, 4, 675-88.

Duffy-Deno, K. and Ebert, R. (1991), 'Public infrastructure and regional economic development: a simultaneous equation approach', *Journal of Urban Economics*, **30**, 329-43.

Dundon-Smith, D. and Gibb, R. (1994), 'The Channel Tunnel and regional economic development', *Journal of Transport Geography*, **2**, 178-89.

Dupuy, G. (1985), *Systèmes, réseaux et territoires*, Paris : Presses de l'Ecole Nationale des Ponts et Chaussées.

Dupuy, G. (1994), 'Réseaux', in Auray J.-P., Bailly A., Derycke P.-H. and Huriot, J.M. (eds), *Encyclopédie d'Economie Spatiale*, Paris: Economica, Bibliothèque de Science Régionale, pp. 145-51.

Dupuy, G. and Stransky, V. (1996), 'Cities and highway networks in Europe', *Journal of Transport Geography*, 107-19.

Eiselt, H. (1992), 'Location modelling in practice', *American Journal of Mathematical and Management Sciences*, **12**, 3-18.

Eiselt, H. and Laporte, G. (1995), 'Objectives in location problems', in Drezner Z. *Facility Location: a Survey of Applications and Methods*, New York: Springer Series in Operations Research, 151-80.

Erkut, E. and Neuman, S. (1989), 'Analytical models for locating undesirable facilities', *European Journal of Operational Research*, **40**, 275-91.

Erkut, E. and Neuman, S. (1992), 'A multiobjective model for locating undesirable facilities', *Annals of Operations Research*, **40**, 209-27.

Erkut, E. and Verter, V. (1995), 'Hazardous materials logistics', in Drezner, Z. (1995), *Facility Location: A Survey of Applications and Methods*, Heidelberg: Springer Verlag, pp. 453-66.

Erlenkotter, D. (1977), 'Facility location with price-sensitive demands: private, public and quasi-public', *Management Science*, **24**, 378-86.

Ferguson, R. (1975), 'Network elongation and the bifurcation ratio', *Area*, **7**, 121-24.

Fisher, H. and Rushton, G. (1979), 'Spatial efficiency of service locations and the regional development process', *Papers of the Regional Science Association*, **42**, 83-97.

Flahaut, B. (1997), *Modèle de localisation et externalités spatiales négatives. Etude d'un cas: la localisation de parcs à conteneurs à l'échelle communale.* Mémoire inédit de licence UCL en Sciences Géographiques. Louvain-la-Neuve.

Flahaut, B., Laurent, M. and Thomas, I. (2001), 'Locating a community recycling center within a residential area: a Belgian case study, *The Professional Geographer* (in press).

Fonseca, M. and Captivo, M. (1993), 'Location of semiobnoxious facilities: some new methods', *Studies in Location Analysis*, **4**, 71-7.

Fotheringham, A., Brundsdon, C. and Charlton, M. (2000), *Quantitative Geography: Perspectives on Spatial Data Analysis*, London: Sage.

Fotheringham, A., Densham, P. and Curtis, A. (1995), 'The zone definition problem in location-allocation modelling', *Geographical Analysis*, **27**(1), 60-77.

Francis, R., Lowe, T., Rushton, G. and Rayco, M. (1999), 'A synthesis of aggregation methods for multifacility location problems: strategies for containing error', *Geographical Analysis*, **31**(1), 67-87.

Francis, R., Lowe, T. and Rayco, M. (1996), 'Row-column aggregation for rectilinear distance *p*-median problems', *Transportation Science*, **30**(2), 160-74.

Francis, R., McGinnis, J. and White, J. (1992), *Facility Layout and Location: An Analytical Approach*, Englewood Cliff, NJ: Prentice-Hall.

Frankhauser, P. (1994), *La fractalité des structures urbaines*, Paris: Anthropos, Collection 'Villes'.

Frankhauser, P. (1997), 'L'approche fractale. Un nouvel outil de réflexion dans l'analyse spatiale des agglomérations urbaines', *Population*, **4**, 1005-40.

Frankhauser, P. and Genre-Grandpierre, C. (1998), 'La géométrie fractale – Un nouvel outil pour évaluer le rôle de la morphologie des réseaux de transport public dans l'organisation spatiale des agglomérations', *Les Cahiers Scientifiques du Transport*, **33**, 41-78.

Fu, K. and Bhargava, B. (1973), 'Tree systems for syntactic pattern recognition', *IEEE Transaction on Computers*, **C-22**(12), 1087-98.

Fujita, M. (1989), *Urban Economic Theory: Land Use and City Size*, Cambridge: Cambridge University Press.

Fujita, M. and Thisse, J.-F. (1996), 'Economics of agglomeration', *Journal of the Japanese and International Economies*, **10**, 339-78.

Furlong, W. and Slotsve, G. (1983), 'Will that be pickup or delivery? An alternative spatial pricing strategy', *Bell Journal of Economics*, **14**, 271-4.

Garrison, W. (1960), 'Connectivity of the interstate highway system', *Regional Science Association, Papers and Proceedings*, **6**, 121-37.

Garrison, W. and Marble, D. (1963), *A Prolegomenon to the Forecasting of Transportation Development,* US Army Technical Report, Northwestern University.

Garvin, W., Crandall, J., John, J. and Spellman, R. (1957), 'Applications of vehicle routing in the oil industry', *Management Science*, **3**, 407-30.

Ginsburgh, V. and Hansen, P. (1974), 'Procedures for the reduction of errors in road network data', *Operational Research Quarterly*, **25**, 321-2.

Giuliano, G. (1989), 'New directions for understanding transportation and land use', *Environment and Planning*, **21A**, 145-59.

Giuliano, G. and Gillepsie, A. (1997), 'Research issues regarding societal change and transport', *Journal of Transport Geography*, **5**(3), 165-76.

Goldman, A. and Witzgall, C. (1970), 'A localization theorem for optimal facility placement', *Transportation Science*, **4**, 406-9.

Goodchild, M. (1979), 'The aggregation problem in location-allocation', *Geographical Analysis*, **11**(3), 240-56.

Goodchild, M. and Mark, D. (1987), 'The fractal nature of geographic phenomena', *Annals of the Association of American Geographers*, **77**, 265-78.

Goovaerts, P. (1997), *Geostatistics for Natural Resources Evaluation*, New York: Oxford University Press.

Gould, P. (1967), 'On the geographical interpretation of eigenvalues?', *Institute of British Geographers Publication*, **42**, 63-92.

Gramlich, E. (1994), 'Infrastructure investment: a review essay', *Journal of Economic Literature*, **32**, 1176-96.

Greenhut, M. (1956), *Plant Location in Theory and Practice*, Chapel Hill: University of North Carolina Press.

Greenhut, M. (1981), 'Spatial pricing in the USA, West Germany and Japan', *Economica*, **48**, 79-86.

Greenhut, M., Norman, G. and Hung, C. (1987), *The Economics of Imperfect Competition*, Cambridge, UK: Cambridge University Press.

Griffith, D. (1996), 'Spatial autocorrelation and eigenfunctions of the geographic weights matrix accompanying geo-referenced data', *Canadian Geographer*, **40**(4), 351-67.

Griffith, D. and Jones, K. (1980), 'Explorations into the relationship between spatial structure and spatial interaction', *Environment and Planning*, **12A**, 187-201.

Griffith, D., O'Neil, M., O'Neil, W., Leifer, L. and Mooney, R. (1986), 'Shape indices: useful measures or red herrings?', *Professional Geographer*, **38**, 263-70.

Gutiérrez, J., Monzon, A. and Pinero, J. (1998), 'Accessibility, network efficiency and transport infrastructure planning', *Environment and Planning*, **30A**, 1337-50.

Haggett, P. and Chorley, R. (1969), *Network Analysis in Geography*, London: Edward Arnold.

Haggett, P. and Chorley, R. (1972), *Network Analysis in Geography*, London: Arnold.

Haggett, P., Cliff, A. and Frey, A. (1977), *Locational Analysis in Human Geography*, London: Arnold.

Haining, R. (1986), 'Spatial models and regional science: a comment on Anselin's paper and research directions', *Journal of Regional Science*, **26**(4), 793-98.

Haining, R. (1993), *Spatial Data Analysis in the Social and Environmental Sciences*, Cambridge: Cambridge University Press.

Hakimi, S. (1964), 'Optimum locations of switching centers and the absolute centers and medians on a graph', *Operations Research*, **12**, 450-9.

Hakimi, S. (1965), 'Optimum distribution of switching centres in a communication network and some related graph theoretic problems', *Operations Research*, **13**, 462-5.

Hakimi, S. (1971), 'Steiner's problem in a graph and its implications, *Networks*, **1**, 113-33.

Hamacher, H., Kmaroth, K., Müller, M., Nickel, S. and Schöbel, A. (1999), *LOLA: Library of Location Algorithms – a Toolkit for Solving Location Problems*, Software of the University of Kaiserslautern, Germany. http://www.mathematik.uni-kl.de/~lola

Handler, G. and Mirchandrani, P. (1979), *Location on Networks*, Cambridge, MA: MIT Press, 233 p.

Handy, S. (1996), 'Methodologies for exploring the link between urban form and travel behavior', *Transportation Research*, **1D**, 151-65.

Hanjoul, P. and Peeters, D. (1985), 'A comparison of two dual-based procedures for solving the *p*-median problem', *European Journal of Operations Research*, **20**, 387-96.

Hanjoul, P., Hansen, P., Peeters, D. and Thisse, J.-F. (1990), 'Uncapacitated location under alternative spatial price policies', *Management Science*, **36**, 41-57.

Hansen, P. and Thisse, J.-F. (1977), 'Multiplant location for profit maximization', *Environment and Planning*, **9A**, 63-73.

Hansen, P., Labbé, M., Peeters, D. and Thisse, J.-F. (1987), 'Facility location analysis', *Fundamentals of Pure and Applied Economics*, **22**, 1-70.

Hansen, P., Peeters, D. and Thisse, J.-F. (1981a), 'On the location of an obnoxious facility', *Sistemi Urbani*, **3**, 299-317.

Hansen, P., Peeters, D. and Thisse, J.-F. (1981b), 'Constrained location and the Weber-Rawls problem', *Annals of Discrete Mathematics*, **11**, 147-64.

Hansen, P., Peeters, D. and Thisse, J.-F. (1983), 'Public facility location models: a selective survey' in Thisse, J.-F. and Zoller, H., *Locational Analysis of Public Facilities*, Amsterdam: North Holland, pp. 223-62.

Hansen, P., Peeters, D. and Thisse, J.-F. (1995), 'The profit-maximizing Weber problem', *Location Science*, **3**, 67-85.

Hansen, P., Peeters, D. and Thisse, J.-F. (1997), 'Facility location under zone pricing', *Journal of Regional Science*, **37**, 1-22.

Hanson, S. and Schwab, M. (1978), 'Accessibility and intraurban travel', *Environment and Planning*, **19A**, 735-48.

Harker, P. (1993), '*Lectures on Computation of Equilibria with Equation-Based Methods*', CORE Lecture Series.

Harris, B. (1993), 'Science in planning: past, present and future', *Journal of the American Planning Association*, **60**, 31-4.

Harvey, D. (1971), 'Social processes, spatial form and the redistribution of real income in an urban system', in Chisholm et al. (eds), *Regional Forecasting*, London: Butterworth, 270-300.

Hay, A. (1975), 'On the choice of methods in the factor analysis of connectivity matrices: a comment', *Institute of British Geographers Publication*, **66**, 163-7.

Haynes, V. (1977), 'The modification of valley patterns by ice-sheet activity', *Geografiska Annaler A*, **59**, 195-207.

Healey, M. and Ilbery, B. (1990), *Location and Change: Perspective on Economic Geography*, Oxford University Press, 381 p.

Henderson, J. (1987), 'System of cities and inter-city trade', in Hansen, P., Labbé, M., Peeters, D., Thisse, H.-F. and Henderson, J. (eds), *System of Cities and Facility Location*, Chur CH: Harwood Academic.

Hodgson, J. and Oppong, J. (1989), 'Some efficiency and equity effects of boundaries in location-allocation models', *Geographical Analysis*, **21**(2), 167-78.

Hodgson, J. and Neuman, J. (1993), 'A GIS approach to eliminating source aggregation error in p-median models', *Location Science*, **1**(2), 155-70.

Hodgson, M., Rosing, K. and Shmulevitz, F. (1993), 'A review of location-allocation applications literature', *Studies in Locational Analysis*, **5**, 3-29.

Hoover, E. (1975), *An Introduction to Regional Economics*, New York: Alfred A. Knopf.

Horn, M. (1996), 'Analysis and computation schemes for p-median heuristics', *Environment and Planning*, **28A**, 1699-708.

Horton, R. (1945), 'Erosional development of streams and their drainage basins: hydrophysical approach to quantitative morphology', *Bulletin of the Geological Society of America*, **56**, 275-370.

Huriot, J, Smith, T. and Thisse J.F. (1989), 'Minimum-cost distances in spatial analysis', *Geographical Analysis*, **2**(4), 294-315.

Huriot, J. and Perreur, J. (1973), 'Modèles de localisation et distance rectilinéaire', *Revue d'Economie Politique*, **83**, 640-62.

Huriot, J. and Perreur, J. (1990), 'Distances, espaces et représentation. Une revue', *Revue d'Economie Régionale et Urbaine*, **2**, 197-237.

Huriot, J. and Perreur, J. (1992), 'Réalité spatiale qualitative et quantification', in Derycke, P.-H. (ed.), *Espace et dynamiques territoriales*, Paris: Economica, Bibliothèque de Science Régionale, pp. 53-92.

Huriot, J. and Thisse J.F. (1985), 'La distance en analyse spatiale: une approche axiomatique', *Document de travail*, 77, Dijon : IME.

Huriot, J., Perreur, J. and Derognat, I. (1994), 'Espace et distance', in Auray J.-P., Bailly, A., Derycke, P.-H. and Huriot, J.M. (eds), *Encyclopédie d'Economie Spatiale*, Paris: Economica, Bibliothèque de Science Régionale, pp. 34–46.

Hurter, A. and. Martinich, J. (1988), *Facility Location and the Theory of Production*, Dordrecht: Kluwer.

Isard, W. (1956), *Location and Space-Economy*, Cambridge, MS: MIT Press.

James, G., Cliff, A., Haggett, P. and Ord, J. (1970), 'Some discrete distributions for graphs with applications to regional transportnetworks', *Geografiska Annaler*, **52B**, 14-21.

Jarvis, R. (1977), 'Spatial variation in ordered tributary basins: a proposed classification', *Geografiska Annaler A*, 241-51.

Jarvis, R. and Werrity, A. (1975), 'Some comments on testing random topology stream network models', *Water Resources Research*, **11**, 309-18.

Johnson, E. (1970), *The Organization of Space in Developing Countries*, Cambridge, MS: Harvard University Press.

Jones, P., Lowe, T., Muller, G., Xu, N., Ye, Y. and Zydiak, J. (1995), 'Specially structured uncapacitated facility location problems', *Operations Research*, **43**, 661-79.

Juel, H. and Love, R. (1985), 'The facility location problem for hyper-rectilinear distances', *AIIE Transactions*, **17**, 94-8.

Kamann D. (1998), 'Modelling networks: a long way to go', *Tijdschrift voor Economische en Sociale Geografie*, **89**(3), 279-97.

Kanemoto, Y. (1980), *Theories of Urban Externalities*, Amsterdam: North-Holland.

Kanemoto, Y. (1987), 'Externalities in space', in Miyao, T. and Kanemoto, Y. (eds), *Urban Dynamics and Urban Externalities*, Chur, CH: Hardwood Academic Publishers.

Kanemoto, Y. and Mera, K. (1985), 'General equilibrium analysis of the benefits of large transportation improvements', *Regional Science and Urban Economics*, **15**, 343-63.

Kansky, K. (1963), 'Structure of transportation networks: relationships between network geography and regional characteristics', University of Chicago, Department of Geography, *Research Papers*, 84.

Kariv, O. and Hakimi, S. (1979), 'An algorithm approach to network location problems II: the *p*-medians', *SIAM Journal of Applied Mathematics*, **37**, 539-60.

Karkazis, J. (1988), 'The general unweighted problem of locating obnoxious facilities on the plane', *Belgian Journal of Operations Research Statistics and Computer Science*, **28**, 43-9.

Karkazis, J. and Boffey, B. (1994), 'Modelling pollution spread', *Studies in Location Analysis*, **7**, 91-104.

Karkazis, J. and Karagiorgis, P. (1987), 'The general problem of locating obnoxious facilities on the plane', Proceedings of the 11th IFORS Conference, Buenos Aires, Brazil.

Kendal, M. (1957), *Multivariate Analysis*, London: Griffin.

King, L. (1984), *Central Place Theory*, Beverly Hills: Sage.

Klein, O. (1998), 'Transport – space – society: from determinism to social and technical innovation. The example of high speed technology', Paper 706 presented at the WTCR in Antwerpen (Belgium).

Klinkenberg, B. (1992), 'Fractals and morphometric measures: is there a relationship?', *Geomorphology*, **5**, 5-20.

Klinkenberg, B. and Goodchild, M. (1992), 'The fractal properties of topography: a comparison of methods', *Earth Surface Processes Landforms*, **17**, 217-34.

Krarup, J. and Pruzan, P. (1990), 'Ingredients of location analysis', in Mirchandrani, P. and Francis, R. (eds), *Discrete Location Theory*, New York: Wiley, 1-54.

Krugman P. (1995), *Development, Geography and Economic Theory*, Cambridge, MA: MIT Press.

Krugman P. (1996), *The Self-Organising Economy*, Cambridge, MA: Blackwell.

Krugman, P. (1991a), *Geography and Trade*, Cambridge MA: MIT Press.

Krugman, P. (1991-b), 'Increasing returns and economic geography', *Journal of Political Economy*, **99**, 483-99.

Krugman, P. (1993), 'The hub effect: or, treeness in interregional trade', in Ethier, W., Helpman, E. and Neary, J. (eds), *Theory, Policy and Dynamics in International Trade: Essays in Honor of Ronald W. Jones*, Cambridge: Cambridge University Press.

Krugman, P. (1998) 'What's new about the new economic geography?', *Oxford Review of Economic Policy*, **14**(2), 7-17.

Krugman, P. and Livas Elizondo, R. (1996), 'Trade policy and the third world metropolis', *Journal of Development Economics*, **49**, 137-50.

Kruskal, J. (1956), 'On the shortest spanning subtree of a graph and the travelling salesman problem', *Proceedings of the American Mathematical Society*, **7**, 48-50.

Labbé, M. and Louveaux, F. (1997), 'Location problems', in Dell'Amico M., Maffioli, F. and Martello, S. (eds), *Annotated Bibliographies in Combinatorial Optimization*, Chichester: Wiley, pp. 261-81.

Labbé, M., Peeters, D. and Thisse, J.-F. (1995), 'Location on networks', in Ball, M., Magnanti, T., Monma, C. and Nemhauser, G. (eds), *Handbooks of Operations Research and Management Science: Networks*, Amsterdam: North-Holland, Elsevier, vol. 8, pp. 551-624.

Labbé, M., Thisse, J.-F. and Wendell, R. (1991), 'Sensitivity analysis in minisum facility location problems', *Operation Research*, **39**, 961-69.

Labys, W. and Yang, C. (1997), 'Spatial price equilibrium as a foundation to unified spatial commodity modelling', *Papers in Regional Science*, **76**(2), 199-228.

Langevin A., Mbagara P. and Campbell, J. (1996), 'Continuous approximation models in freight distribution: an overview', *Transportation Research*, **30B**(3), 163-88.

Laporte, G., Mesa, J. and Ortega, F. (1994), 'Assessing topological configurations for rapid transit networks', Research Paper CRT-999, Montréal, 16 p.

Laurent, M. and Thomas, I. (1999), 'Des places centrales à la localisation optimale', *Revue d'Economie Régionale et Urbaine* (sous presse).

Laurent, M., Peeters, D. and Thomas (1997), 'Spatial externalities and location results: a theoretical case study based on the *SPLP* model', *Studies in Locational Analysis*, **11**, 193-210.

Lawler, E., Lenstra, J., Rinnooy Kan, A. and Shmoys, D. (1985) (eds), *The Travelling Salesman Problem: A Guided Tour of Combinatorial Optimization*, New York: Wiley.

LeBlanc, L., Morlok, E. and Pierskally, W. (1975), 'An efficient approach to solving the road network equilibrium traffic assignment problem', *Transportation Research*, **9**, 309-18.

Lee, D. (1973), 'Requiem for large-scale models', *Journal of the American Institute of Planners*, **39**, 163-78.

Lee, E. (1976), 'An application of fuzzy sets to the classification of geometric figures and chromosome images', *Information Science*, **10**, 95-114.

Lee, R. and Wills, J. (1997), *Geographies of Economies*, London: Arnold.

Liao, K. and Scheidegger, A. (1968), 'A computer model of some branching-type phenomena in hydrology', *Bulletin of the International Association of Hydrological Sciences*, **13**(1), 5-13.

Lienard, P. (1978), *Décibels et indices de bruit*, 2nd edition, Paris: Masson.

List, G. and Mirchandani, P. (1991), 'An integrated network/planar multiobjective model for routing and siting for hazardous material and wastes', *Transportation Science*, **25**, 100-114.

Longley, P. (1998), 'Foundations', in Longley, P., Brooks, S., McDonnell, R. and Macmillan, B. (1998), *Geocomputation: A Primer*, Chichester: John Wiley, pp. 3-15.

Longley, P., Brooks, S., McDonnell, R. and Macmillan, B. (1998), *Geocomputation: A Primer*, Chichester: John Wiley, 278 p.

Lösch, A. (1940), *Die Räumliche Ordnung der Wirtschaft*, Jena: Gustav Fischer. English translation (1954), *The Economics of Location*, New Haven: Yale University Press.

Love, R. (1993), 'Directional bias of the *lp*-norm', *European Journal of Operational Research*, **67**, 287-94.

Love, R. and Morris J. (1972), 'Modelling inter-city road distances by mathematical functions', *Operational Research Quarterly*, **23**(1), 61-71.

Love, R. and Morris, J. (1979), 'Mathematical models of road travel distances', *Management Science*, **25**(2), 130-39.

Love, R. and Morris, J. (1988), 'On estimating road distances by mathematical functions', *European Journal of Operational Research*, **36**, 251-3.

Love, R., Morris J. and Wesolowsky, G. (1988), *Facilities Location: Models and Methods*, Amsterdam: North-Holland.

Love, R., Walker, J. and Tiku, M. (1995), 'Confidence intervals for $l_{k,p,\theta}$ distances', *Transportation Science*, **29**(1), 93-100.

Lowe, J. and Moryades, S. (1975), *The Geography of Movement*, Boston: Houghton Mifflon Company.

Lundberg, U. (1973), 'Emotional and geographical phenomena in psychophysical research', in Downs, R. and Stea, D. (eds), *Image and Environment*, Chicago: Edward Arnold, pp 322-37.

Magnanti, T. and Wong, R. (1984), 'Network design and transportation planning: models and algorithms', *Transportation Science*, **18**, 1-55.

Mandelbrot, B. (1977), *The Fractal Geometry of Nature*, San Francisco: Freeman.

Mark, D. (1975), 'Geomorphometric parameters: a review and evaluation', *Geographical Annaler*, **57A**, 165-77.

Marshall, A. (1890), *Principles of Economics*, London: Macmillan.

Martin, P. (1998), 'Regional policies, growth and geography in Europe', *World Economy*, **21**, 757-74.

Martin, P. and Rogers, C. (1995), 'Industrial location and public infrastructure', *Journal of International Economics*, **39**, 335-51.

Martin, R. (1999), 'The new 'Geographical Turn' in economics: some critical reflections', *Cambridge Journal of Economics*, **23**, 65-91.

Martin, R. and Sunley, P. (1996), 'Paul Krugman's geographical economics and its implications for regional development theory: a critical assessment', *Economic Geography*, **72**(3), 259-92.

Martinez, F. (2000), 'Towards a land-use and transport interaction framework', in Hensher, D. and Button, K. (eds), *Handbook of Transport Modelling*, vol. 1, Oxford: Pergamon, pp. 145-64.

Matthiessen, C. (1993), 'Scandinavian links: Changing the pattern of urban growth and regional air traffic', *Journal of Transport Geography*, **1**, 119-24.

Matzoros, A. and Van Vliet, D. (1992), 'A model of air pollution from road traffic, based on the characteristics of interrupted flow and junction control: Part I - Model description', *Transportation Research*, **26A**(4), 315-30.

Mayeres, I., Ochelen, S. and Proost, S. (1996), 'The marginal external costs of urban transport', *Transportation Research*, **1D**(2), 111-30.

Medda, F., Nijkamp, P. and Rietveld, P. (1998), 'Recognition and classification of urban shapes', *Geographical Analysis*, **30**(3), 304-14.

Mehretu, A., Wittick, R. and Pigozzi, B. (1983), 'Spatial design for basic needs in Eastern upper Volta', *Journal of Developing Areas*, **17**, 383-94.

Melachrinoudis, E. and Cullinane, T. (1985), 'Locating a undesirable facility within a geographical region using the maximin criterion', *Journal of Regional Science*, **25**, 115-27.

Melachrinoudis, E. and Cullinane, T. (1986a), 'Locating an undesirable facility with a minimax criterion', *European Journal of Operations Research*, **24**, 239-46.

Melachrinoudis, E. and Cullinane, T. (1986b), 'Locating a obnoxious facility within a polygonal region', *Annals of Operational Research*, **6**, 137-45.

Melkote, S. (1996), *Integrated Models of Facility Location and Network Design*, PhD Dissertation, Industrial Engineering and Management Sciences, Evanston, Illinois.

Miller, H. (1996), 'GIS and geometric representation in facility location problems', *International Journal of Geographical Information Systems*, **10**, 791-816.

Mills, C. (1970), *The Sociological Imagination*, Harmondsworth: Penguin Books, 256 p.

Ministère du Plan du Niger (1991), *Annuaire Statistique. Séries Longues*. Niamey, Direction de la Statistique et de la Démographie.

Mirchandrani, P. (1990), 'The *p*-median problem and generalizations', in Mirchandrani, P. and Francis, R. (eds), *Discrete Location Theory*, New York: Wiley, 55-117.

Mirchandrani, P. and Francis, R. (eds) (1990), *Discrete Location Theory*, New York: Wiley.

Mishan, E. (1971), 'The postwar literature on externalities: an interpretative essay', *Journal of Economic Literature*, **9**, 1-18.

Montgomery, K. (1996), 'Sinuosity and fractal dimension of meandering rivers', *Area*, **28**(4), 491-500.

Morris, J., Dumble, P. and Wigan, M. (1979), 'Accessibility indicators for transport planning', *Transportation Research*, **13A**(1), 91-109.

Muller, J. (1982a), 'Non-Euclidean geographic spaces: mapping functional distances', *Geographical Analysis*, **14**(3), 189-203.

Muller, J. (1982b), 'La cartographie d'une métrique non euclidienne: les distances-temps', *L'Espace Géographique*, **3**, 215-27.

Mulligan, G. (1984), 'Agglomeration and central place theory: a review of the literature', *International Regional Science Review*, **9**, 1-42.

Mulligan, G. (1996), 'Myopic spatial competition: boundary effects and network solutions', *Papers in Regional Science*, **75**, 155-176.

Mun, S. (1997), 'Transport network and system of cities', *Journal of Urban Economics*, **42**, 205-21.

Munnel, A. (1992), 'Infrastructure investment and economic growth', *Journal of Economic Perspectives*, **6**, 189-98.

Murayama, Y. (1994), 'The impact of railways on accessibility in the Japanese urban system', *Journal of Transport Geography*, **2**, 87-100.

Murray, A., Church R., Gerrard, R. and Tsui, W. (1998), 'Impact models for siting undesirable facilities', *Papers in Regional Science*, **77**(1), 19-36.

Musso, A. and Vuchic, V. (1988), 'Characteristic of metro networks and nethodology for their evaluation', *Transportation Research Record*, **1162**, 22-33.

Nagurney, A. (1993), *Network Economics: A Variational Inequality Approach*, Dordrecht: Kluwer Academic Publishers.

Nagurney, A. and Zhang, D. (1996), *Projected Dynamical Systems and Variational Inequalities with Applications*, Dordrecht: Kluwer Academic Publishers, 296 p.

Nagurney, A. and Zhang, D. (1997a), 'Formulation, stability, and computation of traffic network equilibria as projected dynamical systems', *Journal of Optimization Theory and Applications*, **93**(2), 417-44.

Nagurney, A. and Zhang, D. (1997b), 'Massively parallel computation of dynamic traffic networks modeled as projected dynamical systems', *Network Optimization*, **450**, 374-96.

Nagurney, A. and Zhao, L. (1993), 'Variational-inequalities and networks in the formulation and computation of market equilibria and disequilibria: the case of direct demand functions', *Transportation Science*, **27**(1), 4-15.

Nakicenovic, N. (1995), 'Overland transportation networks: history of development and future prospects', in Batten, D., Casti, J. and Thord, R. (eds), *Networks in Action: Communication, Economics and Human Knowledge*, Berlin: Springer-Verlag, pp. 195-228.

Nijkamp, P., Perrels, A. and Schippers, L. (1995), 'The strategic role of new infrastructure networks in Europe', in Batten, D., Casti, J. and Thord, R. (eds), *Networks in Action: Communication, Economics and Human Knowledge*, Berlin: Springer-Verlag, pp. 229-50.

Nijkamp, P., Rietveld, P. and Salomon, I. (1990), 'Barriers in spatial interactions and communications', *Annals of Regional Science*, **24**, 237-52.

Nutley, S. (1984), 'Planning for rural accessibility provision: welfare, economy and equity', *Environment and Planning*, **16A**, 357-76.

Offner, J. (1993), 'Les effets structurants du transport: mythe politique, mystification scientifique', *L'Espace Géographique*, **3**, 233-42.

Offner, J. (1997), 'Les effets structurants du transport: mythe politique, mystification scientifique',Paris : Ponts Formation Edition, January, internal report.

Okafor, F. (1990), 'The spatial dimensions of accessibility to general hospitals in rural Nigeria', *Socio-Economic and Planning Sciences*, **24**, 295-306.

Openshaw, S. and Taylor, P. (1979), 'A million or so correlation coefficients: three experiments on the modifiable areal problem', in Wrigley, N. (ed.), *Statistical Applications in the Spatial Sciences*, London: Pion, 127-44.

Oppong, J. (1996), 'Accommodating the rainy season in Third World location-allocation applications', *Socio-Economic and Planning Sciences*, **30**(2), 121-37.

Oppong, J. and Hodgson, J. (1994), 'Spatial accessibility to health care facilities in Suhum District, Ghana', *Professional Geographer*, **46**(2), 199-209.

Ord, J. (1967), 'On a system of discrete distributions?', *Biometrika*, **54**, 649-56.

Ottaviano, G. and Thisse, J.-F. (1998), 'Agglomeration and trade revisited', *CEPR Discussion Paper* 1903.

Ottaviano, G.P. and Puga, D. (1998), 'Agglomeration in the global economy: a survey of the 'new geography', *World Economy*, **21**(6), 707-31.

Papadimitriou, C. and Karkazis, J. (1991), 'A quantitative model for evaluating the impact of a pollution stack', *Yugoslav Journal of Operations Research*, **1**, 27-44.

Papageorgiou, Y. (1978), 'Spatial externalities I: theory, II: applications', *Annals of the Association of American Geographers*, **68**(4), 465-92.

Papini, P. (1994), 'How can we locate centers?', *Studies in Locational Analysis*, **7**, 143-52.

Patel, N. (1979), 'Locating rural social service centres in India', *Management Science*, **25**, 22-30.

Pearce, C. (1974), 'Locating concentric ring roads in city', *Transportation Science*, **8**, 142-68.

Peeters, D. (1995), *Contribution to the Theory of the Location of the Monopolistic Firm*, Thèse d'Agrégation de l'Enseignement Supérieur en Sciences Géographiques, U.C.L., Louvain-la-Neuve (unpublished).

Peeters, D. and Thisse, J.-F. (1990), 'Spatial price policies and the location of the firm', in Chatterji, M. and Kuenne, R. (eds), *New Frontiers in Regional Science*, London: MacMillan, 57-74.

Peeters, D. and Thisse, J.-F. (1995), 'Economic models of firm location: theory and algorithms', in Drezner, Z. (ed.), *Location Theory*, Heidelberg: Springer-Verlag, pp. 331-66.

Peeters, D. and Thisse, J.-F. (1996), 'Zone pricing', *Journal of Regional Science*, **36**, 291-301.

Peeters, D. and Thomas, I. (1995), 'The effect of the spatial structure on location-allocation results', *Transportation Science*, **29**(4), 366-73.

Peeters, D. and Thomas, I. (1997) 'Distance-*lp* et localisations optimales. Simulations sur un semis aléatoire de points', *Cahiers Scientifiques du Transport*, **31**, 55-70

Peeters, D. and Thomas, I. (2000), 'Choosing a measure of distance for applied location-allocation models: simulations based on the *lp*-distance and the *k*-median Model', *Geographical Systems*, 2: 167-84.

Peeters, D., Thisse, J.-F. and Thomas, I. (1998a), '*Modèles opérationnels de localisation des équipements collectifs*', Congrès des Economistes Belges de Langue Française, Charleroi, novembre 1998. Commission 2: Localisation des activités économiques: efficacité versus équité, pp. 135-58.

Peeters, D., Thisse, J.-F. and Thomas, I. (1998b), 'Transportation networks and the location of human activities', *Geographical Analysis*, **30**(4), 355-71.

Peeters, D., Thisse, J.-F. and Thomas, I. (2001), 'High speed connections ans the location of human activities', *Environment and Planning A*, **32**(12), 2097-112.

Perreur, J. (1989), 'L'évolution des représentations de la distance et l'aménagement du territoire', *Revue d'Economie Régionale et Urbaine*, **1**, 115-41.

Perreur, J. and Thisse, J.-F. (1974), 'Central metrics and optimal locations', *Journal of Regional Science*, **14**, 411-21.

Petrakos, G. (1997), 'The regional structure of Albania, Bulgaria and Greece: implications for a cross-border cooperation and development', *European Urban and Regional Studies*, **4**, 195-210.

Peuquet, D. (1988), 'Representations of geographic space: toward a conceptual synthesis', *Annals of the Association of the American Geographers*, **78**(3), 375-94.

Peura, M. (1997), 'A statistical classification method for hierarchical irregular objects', in de Bimbo, A. (ed.), *Image Analysis and Processing*, 9th International Conference (ICIAP97), vol. 1310 of *Lecture Notes in Computer Science*, pp. 604-11.

Peura, M. (1998), 'The self-organizing map on trees', *Neural Processing Letters*, **8**(2), 155-162.

Phlips, L. (1983), *The Economics of Price Discrimination*, Cambridge, U.K.: Cambridge University Press.

Pigou, A. (1920), *The Economics of Welfare*, London: Macmillan.

Pitts, F. (1965), 'A graph theoretic approach to historical geography', *Professional Geographer*, **17**, 15-20.

Plassard, F. (1976), *Les autoroutes et le développement régional*, Paris : Economica.

Plastria, F. (1995), 'What is the best aggregation point? A case for the gravity centre', *Report BEIF/88, Brussels, V.U.B., Centrum voor Bedrijfsinformatica*, 14 p.

Plummer, P. (1996a), 'Spatial competition amongst hierarchically organized corporations: prices, profits, and shipment patterns', *Environment and Planning*, **28A**, 199-222.

Plummer, P. (1996b), 'Competitive dynamics in hierarchically organized markets: spatial duopoly and demand asymmetrics', *Environment and Planning*, **28A**, 2021-40.

Plummer, P., Haining, R. and Sheppard, E. (1998), 'Spatial pricing in interdependent markets: testing assumptions and modelling price variation. A case study of gasoline retailing in St Cloud, Minnesota', *Environment and Planning*, **30A**, 67-84.

Pooler, J. (1995), 'The use of spatial separation in the measurement of transportation accessibility', *Transportation Research*, **29A**(6), 421-27.

Quinet, E. (1992), *Infrastructures de transport et croissance*, Paris: Economica.

Quinet, E. (1998), *La grande vitesse, une chance pour le développement économique*, Eurailspeed 98, Berlin.

Raffestin, C. (1981), *Pour une géographie du pouvoir*, Paris: Litec.

Raghavan, S. and Magnanti, T. (1997), 'Network connectivity', in Dell'Amico, M., Maffioli, F. and Martello, S. (eds), *Annotated Bibliographies in Combinatorial Optimization*, Chichester: Wiley, pp. 335-54.

Ratick, S. and White, A. (1988), 'A risk-sharing model for locating obnoxious facilities', *Environment and Planning*, **15B**, 165-79.

Rephann, T. and Isserman, A. (1994), 'New highways as economic development tools: an evaluation using quasi-experimental matching methods', *Regional Science and Urban Economics*, **24**, 723-51.

ReVelle, C., Cohon, J. and Shobrys, D. (1991), 'Simultaneous siting and routing in the disposal of hazardous wastes', *Transportation Science*, **25**, 138-45.

Richard, D., Beguin, H. and Peeters, D. (1990), 'Fire stations location in a rural environment', *Environment and Planning*, **22A**, 39-52.

Richardson, A. and Young, W. (1982), 'A measure of linked-trip accessibility', *Transportation Planning and Technology*, **7**, 73-82.

Rienstra, S., Rietveld, P., Hilfering, M. and Bruinsma, F. (1998), 'Road infrastructure and corridor development', in Lundqvist, L., Mattson, L., Kim, T. (eds), *Network infrastructure and the Urban Environment*, Berlin: Springer-Verlag, pp. 395-414.

Rietveld, P. (1989a), 'Infrastructure and regional development. A survey of multiregional economic models', *Annals of Regional Science*, **23**, 255-74.

Rietveld, P. (1989b), 'Employment effects of changes in transportation infrastructure: methodological aspects of the gravity model', *Papers of the Regional Science Association*, **66**, 19-30.

Rietveld, P. (1993), 'Methods for planning urban facilities in rural areas of developing countries: a comparison of approaches', *International Regional Science Review*, **15**(3), 345-56.

Rietveld, P. (1994), 'Spatial economic impacts of transport infrastructure supply', *Transportation Research*, **28-A**(4), 329-41.

Rietveld, P. and Wintershoven, P. (1998), 'Border effects and spatial autocorrelation in the supply of network infrastructure', *Papers in Regional Science*, **77**(3), 265-76.

Rondinelli, D. (1990), 'Locational planning and regional economic development: appropriate methods in developing countries', *International Regional Science Review*, **13**, 241-48.

Rosing, K., Hillsman, E. and Rosing-Vogelaar, H. (1979), 'A note comparing optimal and heuristic solutions to the *p*-median problem', *Geographical Analysis*, **11**, 87-9.

Roy, J. (1999), 'Areas, nodes and networks: some analytical considerations', *Papers in Regional Science*, **78**(2);135-56.

Ruhigira, J. (1994), *Effets de l'agrégation de la demande sur les configurations optimales de l'offre générées par le modèle de localisation-affectation de la p-médiane*, Thèse de doctorat en sciences, UCL, Louvain-la-Neuve, 27-01-94.

Rushton, G. (1984), 'Use of location-allocation models for improving the geographical accessibility of rural services in developing countries', *International Regional Science Review*, **9**(3), 217-40.

Rushton, G. (1988), 'Location theory, location-allocation models and service development planning in the Third World', *Economic Geography*, **64**, 97-120.

Rushton, G. (1989), 'Applications of location models', *Annals of Operations Research*, **18**, 25-42.

Saddala, E. and Magel, S. (1980a), 'The perception of traversed distance', *Environment and Behavior*, **12**, 65-79.

Saddala, E. and Staplin, L. (1980b), 'The perception of traversed distance, intersections', *Environment and Behavior*, **12**, 167-82.

Sakakibara, K., Matsuyama, K. and Suzuki, A. (1996), 'Error free L_1-Voronoi diagram construction algorithm and its application to *p*-median problem of L_1 distance in a unit square', *Studies in Locational Analysis*, **9**, 113-16.

Sasaki, K. (1992), 'Trade and migration in a two-city model of transportation investments', *Annals of Regional Science*, **26**, 305-17.

Schärlig, A. (1969), *Localisation optimale et théorie des graphes*, Genève: Librairie Droz, 156 p.

Scheidegger, A. (1965), 'The algebra of streamorder numbers', *US Geological Survey Professional Paper*, **525b**, B187-B189.

Shasha, D., Wang, J., Zhang, K. and Shih, F. (1994), 'Exact and approximate algorithms for unordered tree matching', *IEEE Transactions on Systems, Man and Cybernetics*, **24**(4), 668-78.

Shreve, R. (1966), 'Statistical low of stream numbers', *Journal of Geology*, **74**, 17-37.

Shreve, R. (1967), 'Infinite topologically random channel networks', *Journal of Geology*, 178-86.

Smith, D. (1975), *Patterns in Human Geography: an Introduction to Numerical Methods*, New York: Newton Abbot, David and Charles.

Smith, D. (1977), *Human Geography: a Welfare Approach*, London: Edward Arnold.

Soguel, N. (1994), *Evaluation monétaire des atteintes à l'environnement: une étude hédoniste et contingente sur l'impact des transports*, Neuchâtel : EDES.

Stock, R. (1982), 'Distance and utilisation of health facilities in rural Nigeria', *Social Sciences and Medicine*, **17**, 563-50.

Stowers, C. and Palekar, V. (1993), 'Location models with routing considerations for a single obnoxious facility', *Transportation Science*, **27**(4), 350-62.

Strahler, A. (1964), 'Quantitative geomorphology of drainage basins and channel networks', in Chow, V. (ed.), *Handbook of Applied Hydrology*, New York: McGraw-Hill, 4.40-4.74.

Suarez-Villa, L., Gaoutzi, M. and Stratigea, A. (1992), 'Territorial and border barriers in information and communication networks: a conceptual exploration', *Tijdschrift voor Economische en Sociale Geografie*, **83**, 93-104.

Taaffe, E. (1974), 'The spatial view in context', *Annals of the Association of American Geographers*, **64**, 1-16.

Taaffe, E., Gauthier, H. and O'Kelly, M. (1996), *Geography of Transportation*, New York: Englewood Cliffs:Prentice Hall, 422 p.

Teitz, M. and Bart, P. (1968), 'Heuristic methods for estimating the generalized vertex median of a weighted graph', *Operations Research*, **16**, 955-61.

Tewari, V. (1992), 'Improving access to services and facilities in developing countries', *Journal of Transport Geography*, **5**, 127-36.

Thisse, J.-F. (1975), *Contribution à la théorie microéconomique spatiale*, Thèse de doctorat en Sciences Economiques, Faculté de Droit, Université de Liège, 175 p.

Thisse, J.-F. (1994), 'L'équité spatiale', in Auray, J.-P., Bailly, A., Derycke, P.-H. and Huriot, J.M. (eds), *Encyclopédie d'Economie Spatiale*, Paris: Economica, Bibliothèque de Science Régionale, pp. 225-31.

Thomas, I. (1993), 'Sensibilité du découpage spatial optimal des services d'urgence à la définition de la demande. Quelques réflexions sur un exemple mésogéographique', *L'Espace Géographique*, **4**, 318-22.

Thomas, I. (1994), *Localisation optimale des permanences mobiles de la Gendarmerie*, Bruxelles: Etat-Major de la Gendarmerie (report associated to research contract).

Tinkler, K. (1972), 'The physical interpretation of eigenvalues of dichotomous matrices', *Institute of British Geographers Publication*, **55**, 17-46.

Tinkler, K. (1975), 'On the choice of methods in the factor analysis of connectivity matrices: a reply', *Institute of British Geographers Publication*, **66**, 168-71.

Tinkler, K. (1977), *An Introduction to Graph Theoretical Methods in Geography*, Norwich: Geo-abstracts.

Tinkler, K. (1979), 'Graph theory', *Progress in Human Geography*, **3**, 85-116.

Tomkins, J., Topham, N., Twomey, J. and Ward, J. (1998), 'Noise versus access: the impact of an airport in an urban property market', *Urban Studies*, **35**, p. 243-58.

Troin, J. (1995), *Rail et aménagement du territoire. Des héritages aux nouveaux défis*, Aix-en-Provence: Edisud, 263 p.

Valeo, C., Baetz, B. and Tsanis, I. (1998), 'Location of recycling depots with GIS', *Journal of Urban Planning and Development-Asce*, **124**(2), 93-9.

van den Berg, L. and Pol, P. (1998), 'The urban implications of the developing european high-speed-train network', *Environment and Planning*, **16C**(4), 483-97.

Vasconcellos, E. (1997), 'Rural transport and access to education in developing countries: policy issues', *Journal of Transport Geography*, **5**, 127-36.

Venables A. (1996), 'Equilibrium locations of vertically linked industries', *International Economic Review*, **37**, 341-59.

Verhoef, E. (1994a), 'Efficiency and equity in externalities: a partial equilibrium analysis', *Environment and Planning*, **26A**, 361-82.

Verhoef, E. (1994b), 'External effects and social costs of road transport', *Transportation Research*, **28A**, 273-87.

Vickerman, R. (1987), 'The Channel Tunnel: consequences for regional growth and development', *Regional Studies*, **21**, 187-97.

Vickerman, R. (1994-a), 'The Channel Tunnel and regional development in Europe: an overview', *Applied Geography*, **14**, 9-25.

Vickerman, R. (1994-b), 'Transport infrastructure and region building in the European Community', *Journal of Common Market Studies*, **32**, 1-24.

Vickerman, R. (1995), 'The regional impacts of Trans-European networks', *Annals of Regional Science*, **29**, 237-54.

Wackernagel, H. (1998), *Multivariate Geostatistics*, Berlin: Springer.

Ward, J. and Wendell, R. (1980), 'A new norm for measuring distance which yields linear location problems', *Operations Research*, **28**, 836-44.

Weber, A. (1909), *Theory of the Location of Industries*, Chicago: University of Chicago.

Wegener, M. (1994), 'Operational urban models: state of the art', *Journal of the American Planning Association*, **60**, 17-29.

Wegener, M. (1998), 'Applied models of urban land use, transport and environment: state of the art and future developments', in Lundqvist, L., Mattson, L. and Kim, T. (eds), *Network infrastructure and the Urban Environment*, Berlin: Springer-Verlag, pp. 245-67.

Werner C. and Smart, J. (1973), 'Some new methods of topologic classification of channel networks', *Geographical Analysis*, **5**, 271-95.

Werrity, A. (1967), 'The topology of stream networks', in Chorley, R. (ed.), *Spatial Analysis in Geomorphology*, London: Methuen, 167-96.

Wesolovsky, G. and Love, R. (1971), 'The optimal location of new facilities using rectangular distances', *Operations Research*, **19**, 124-34.

Wheeler, J. and Muller, P. (1986), *Economic Geography*, New York: Wiley, 412 p.

Wilson, R. (1996), *Introduction to Graph Theory*, Harlow: Longman.

Wrigley, N. (1985), *Categorical Data Analysis for Geographers and Environmental Scientists*, London: Longman.

Yinger, J. (1993), 'Around the block: urban models with a street grid', *Journal of Urban Economics*, **33**, 305-30.

Index

accessibility
  community recycling centres 255
  health care facilities 217-18, 222, 233-34
asymmetric networks
  hexagonal lattices 61, 62
  market area shapes 66
  optimal location sensitivity 65-66
autarky, case of 21, 40
  allocation 53
  distance
    average 49-51, 59-60, 63-64, 64, 71
    frequency 58-59, 65, 71
    maximum 52-53
  k-median model 47-48
    network simulations 45-47, 49
    outputs 63-70
    results 49-61
  market area shape 55-57, 66-70, 73
  network designs used 61-63
  network shape, influence of 44-45, 60-61
  optimal location sensitivity 65-66
  ring road location 70-71, 73, 80
  supply site location 54-55

borders, common market
  barriers, removing 102-3, 119
  context 101-5
  gates
    length 115-18, 119
    location 112-14, 119
    numbers 107-12
  interregional high speed connections 119-34
    equal density economies 121-30
    experiment description 120-21
    facility numbers 109-10, 112-13, 115, 126-27
    intermediate stops 127-28
    optimal locations 110-18, 121-25, 128-30
    unequal density economies 131-34
  methodology 105-7
    limitations of 135-36
  permeability 106-7, 119
  policy implications 135-36, 253
  toy-networks used 107-9, 120-21
  transport infrastructure 103-5

capacity constraints 73
central place theory 121, 122
Channel Tunnel 107
Christaller-Lösch theory 61, 67, 75, 78, 105
circumradial networks 46, 49, 76, 77, 181-82
  deconcentration 93
  facility numbers 82-83, 85, 86
  optimal locations 83-84, 85
  price policies 182-87
  ring road location 53, 70-71, 73, 80, 253
common markets 22, 40
  demand
    non-uniform 82-83, 84-87
    uniform 82-84
  facility numbers 77, 82-83, 85, 86, 87, 88
  fixed costs 82-83, 84-87
  lattice networks 79-92, 102
  legacy of past 252-53
  network density 81-82, 89-92
  network shape 252
  optimal locations 83-84, 85
    integrated networks 87, 96-97
    separated networks 87, 98-99
  peripheral ring road, impact of 253
  radial and grid economies 77-78
    merging 87-89
  SPLP model 75-76, 78-79

transport infrastructure 76–77, 94
transportation/pollution costs trade-off 255
see also borders, common market
community recycling centre
accessibility 255
demand, measure of 247–50
distance 245
inputs 238–41
nuisance, and distance 238–39, 242–44
obnoxious facilities, optimisation criteria 236–37
optimal locations 242–44, 245–46
pollution spread 240–41, 243
results
several facilities 245–47
single facility 242–44
studied environment (La Bruyère) 241
transportation/pollution costs trade-off 236–38
user characteristics, of recycling centres 247–48
compact city *versus* suburbanisation process 2
congestion 3, 73, 77, 95, 105

delivery pricing policy (*m*) 177, 179–80, 182–87, 187–89
demand
common markets 82–97
health care facilities case study 220, 221
measure of 247–50
price policies 184–87
random *versus* pyramidal distribution 84–85
spatial distribution 257–58
density *see* networks, density
developing countries
accessibility 217–18, 222, 233–34
public sector planning 217
radial networks 94–95
trade policy 95
transportation infrastructure 76–77, 94
see also health care facilities case study
discriminatory pricing policy (*D*) 175–76, 178–81, 182–87, 187–88
distance
evaluating 139–40
measurement 8, 9–10, 43–44, 144
minimisation 222
and nuisance 197–98, 238–39
perception 144
spatial interdependence 139
see also Euclidean networks; Euclidean space
distance-prediction function
distance
average and maximum 151–52, 156–57, 161–63, 168
distribution 153, 154–55, 160, 161
measuring, error-prone nature of 144
equity efficiency ratio 156–57, 161, 167–68
experiment design 145–50
lattice irregularity 149, 162–66
l_p-distance 144–69
l_p-norm 143-4, 146, 169
optimal locations 153, 154, 157–59, 163–66, 168
parameters, estimating 22–23, 169
power transformation effect 167–68
practical implications 254
reference axis orientation effect 148–49, 157–57, 160–62
settlement system 145

economic geography 15, 103, 105, 171
and economics 16–17, 18
location-allocation techniques 20–21
production/transportation costs trade-off 19–20
roots 17–19
supply-driven approach 20
edges 2, 7, 10, 12
defining 28, 30
market area shape 67
measurement of 22
see also links
efficiency–equity balance 2, 9, 52, 63, 70–71, 156–57, 228
eigenvalues 33, 34
elementary cells 45, 145
equity–efficiency balance 2, 9, 52, 63, 70–71, 156–57, 161, 167–68, 228
EU *see* European Union

Euclidean networks 46, 181-82
  allocation 53
  distance
    average 49-51
    frequency distribution 58
    maximum 52-53
  hexagonal lattices 61, 62
  market area shapes 55, 66
  supply site location 54
Euclidean space 46, 49, 50, 140, 146-47, 148, 157, 161, 165, 181, 196, 222
  geometrical *versus* topological space 72
European Union 77, 102
  HST (high speed railway) 102-3, 104 (*see also* interregional high speed connections)
  transport barriers, removing 102, 107-19
  see also borders, common market; common markets
externalities 140-41, 196-98, 211-12
  defining 196-97
  distance, average 201-5
  as indirect effect 196-97
  k-median model, modified 199-201
  monetary value of, estimating 239-40
  noxious facilities 194
  obnoxious facilities
    benefits and disbenefits 199
    disutilities accompanying 211-12
    nuisance and distance 197-98
  optimal locations, modification of 206-10
  pollution spread 195
  settlement system 195-96
  as spatial 197
  transportation/pollution costs trade-off 193, 199-201, 202, 204-5, 211-12
  see also community recycling centre case study

facility location analysis 21, 22
  network shape 39
  problem formulation 43
  production/transportation costs trade-off 75-76, 93, 105, 127, 135
facility workload 66, 70, 71
Filingué (Niger) *see* health care facilities case study

garbage management, problem of 235
gates 102-3
  common market borders 107-19
  national networks 95
  permeability of 106-7
gateway cities 111, 115, 135, 254
  bottleneck cities 121-22, 127-28
geographical perspective
  GIS and geocomputation 15-16
  spatial economic analysis 17
  spatial modelling 13-15
  see also economic geography
geomorphometry 33, 34
GIS (geographical information systems) 16
  distance-predicting function 22-23
  and spatial modelling 15
  technologies 13, 234, 254, 257
globalisation 77, 95
graphs 38
  classifying 27
  defining 28, 30
  directed (digraph) 30
  graph theory, terminology of 30-31, 32, 32
  isomorphism 31
  transportation networks 11, 12
  see also edges; links; networks; nodes
grid networks
  common market borders 106
  dispersed pattern of production 93
  flexibility of location choices 108
  versus radial networks 77-78, 87-89, 252
  see also rectilinear networks

Hakimi theorem 39, 44
health care facilities case study 23, 215, 217-34
  accessibility restrictions 228, 232, 233-34
  climate 220
  data limitations 218
  demand pattern 220, 221

distance
average weighted 226–27, 229, 230
maximum 227
variation 224–25
hierarchy 232, 233
network 220–21
enhancement 221, 223–24, 230–32
shape, influence of 234
optimal locations 225–26, 227–29, 230–31, 232
seasonal variations 218, 220, 222–23, 223, 226–29, 233, 252
hedonist prices method 239–40
hexagonal lattices 60–61, 71–72
radial networks 61–63
HST *see* TGV

indexing *see* networks, indexing
industrial belt (US) 93
intercitytransport investment models 4
interregional communication networks 101–2
interregional high speed connections 119–34
equal density economies 121–30
no TGV 121–25
one TGV 126–30
experiment description 120–21
facility numbers 109–10, 112–13, 115, 126–27
intermediate stops 127–28
optimal locations 110–18, 121–25, 128–30
policy recommendations 253–54
unequal density economies 131–34

k-median model 6, 9, 21–22, 40, 172
autarky, case of 47–48
community recycling centre case study 238
distance, optimal average 201–5
distance-prediction function 143, 150
externalities 199–201
health care facilities case study 215, 222–24
network morphology 72
sensitivity analyses 43
toy-networks 44, 47–48

La Bruyère (Belgium) *see* community recycling centre case study
land-use planning 2–4
gate length 119
network morphology 44
policy implications 251–56
transportation models 14
transportation/pollution costs trade-off 250
lattice networks 79-92, 102, 108, 120–21, 145, 181, 191, 192, 195–96
irregularity 149, 162–66
links 28
and average distance 50, 51, 63, 64
impact of additional 95
transportation networks 10–11
types of 28–29
weights, influence of 140
location-allocation models
computational capability 8–9
formulation approaches 7–8
inputs 7, 9–10, 12
outputs 7, 40
purpose of 6
recent developments 6–7
shortest-path matrix 30–31
see also *k*-median model; SPLP; UFLP
location theory 6, 12, 18
network morphology 104
lock-in effects 93–94
l_p-distance 150–51, 155
l_p-norm 144, 148
k-median model 143, 150
see also distance-prediction function

Marshallian agglomeration economies 18–19
methodology 73
limitations of 135–36, 251–52
Minkowsky distance (l_p-norm) 146, 147
mixed network 81
mixed pricing policy (*m*) 177, 179-80, 182–87, 187–89
multidisciplinarity, need for 259

NAFTA 77
NDC (network design problems) 12
networks
analysis 12

density 63, 65, 69, 70, 71-72
borders 107-8
common markets 89-92
dense *versus* sparse networks 120, 131-34
versus shape 78
design models
mimimum spanning tree model 11, 12
operations research models 11
Steiner tree problem 11, 12
design problems (NDC) 12
indexing
eigenvalues 33, 34
fractals 33, 34-35
geomorphometry 33, 34
graph theory 32, 32
neural modelling techniques 35
pattern recognition 35
shortest path matrix 33-34, 33
synthesis 21, 27, 31, 36, 251
transportation access level 35-36
morphology
activity location, impact on 27, 31
functional and spatial defined 30
k-median problem 72
shape
autarky, case of 44-45, 60-61
and location-allocation results 21, 22, 39-41, 44, 169, 215, 232, 233
measuring 27-31
price policies 172, 255
transportation infrastructure 39
structure
linkage types 28-29
matrices 30-31
topology, and facility location 12-13
types
and allocation 53
and average distance 49-50, 51, 59-60
and maximum distance 52-53
urban level 3
neural modelling techniques 35
Niger (Africa) *see* health care facilities case study
NIMBY (not in my back yard) phenomenon (community recycling centres) 236
nodes
human geography 10
network 2, 7
defining 27-28, 28, 30
Steiner nodes 12
noise pollution 24
estimating external costs of 239-40
nuisance, and distance 197-98, 238-39

order, notion of 34

Pareto efficiency 196
permeability, of borders 106-7, 119
pick-up pricing policy (*m*) 177, 179-80, 182-87, 187-89
pollution
costs 140-41
as negative externality 23
spread, modelling 240-41
see also community recycling centre case study; externalities; transportation costs/pollution costs trade-off
polynomial alogorithms 8-9
positivism, and scientific method 14
price location problem (PLP) 173-75
price policies 23, 140, 171-92, 173-81, 175-78
circumradial networks 182-87
demand distribution 184-87, 189-90
discriminatory pricing policy (*D*) 175-76, 178-81, 182-87, 187-88
gasoline stations, accessibility of 192
integrated price-production-location models 172
market area comparison 178-81
mixed (pick-up or delivery) pricing policy (*m*) 177, 179-80, 182-87, 187-89
network shape 172, 185, 188, 255
networks used 181-82
optimal locations 182, 183, 184, 187-88, 191
price location problem (PLP) 173-75
simulation constraints, alleviating 191
spatial pricing 171, 172
uniform delivered pricing policy (*U*) 176, 178-80, 182-87, 187-90
uniform mill pricing policy (*M*) 178-80, 182-87, 187-90

zone pricing policy (Z) 177–78, 180–81, 182–87, 187–88
production costs/transportation costs trade-off
economic geography 19–20
facility location analysis 75–76, 93, 105, 127, 135
SPLP (simple plant location problem) 75–76, 78–79, 92–93, 105, 106–7
public facilities
equity–efficiency problem 9
production/transportation costs trade-off 105
'putty-clay geography' 93

quadratic assignment problem 6

radial networks 46, 47, 181–82
allocation 53
centralised policies 76
developing countries 94–95
distance
average 49–51, 63
frequency distribution 58–59
maximum 52–53
facility numbers 77, 82–83, 85, 86
and grid networks, integration of 87–89
hexagonal lattices 61–63
irregular spatial organisation of 72
market area shapes 55–57, 66–69
mixed network 81
optimal locations 65–66, 83, 84, 85
production concentration 93, 252
supply site location 55
railway networks *see* interregional high speed connections
real-world problems *versus* toy-networks 79, 215–16
rectangular networks *see* rectilinear networks
rectilinear networks 80, 181–82
decentralised policies 76
facility numbers 77, 82–83, 85, 86
optimal locations 83, 84, 85
regional development
impact of public policies on 171
transportation costs 3
transportation infrastructure policy 101–5
regular networks 49–50
ring road location 53, 70–71, 73, 80, 253

scientific method 14
sensitivity analyses 4, 9, 72
autarky, case of 43
computational tractability/model accuracy trade-off 143
service supply 44
shape *see* networks, shape
shortest path matrix 33–34, 33
simple plant location problem *see* SPLP
societal change, in context of research 259
spatial analysis
edges, time-consuming measurement of 22
methods 13–14
spatial economics 93–94
location theory 172
spatial competition 172
spatial interaction models 14
spherical distance 146, 147
SPLP (simple plant location problem) 20, 21, 22, 40, 73, 75–78, 172
approximation of market solution 106
common markets 75–76, 78–79, 82–92
demand distribution 93–94
interregional high-speed connection experiment 120–21
production/transportation costs trade-off 75–76, 78–79, 92–93, 105, 106–7
reasons for using 76, 105–6
socially optimum facility configuration 105–6
statistical methods 13–14, 257

TGV (France) 101, 104, 106
intermediate stops 127
link experiment 120–21
see also interregional high speed connections
Third World countries *see* developing countries
topology, network 10, 12–13
toy-networks 3, 4, 10, 21, 22, 40–41, 45–49, 135

common market borders 107-9, 120-21
limitations of 257
network types 45-47
versus real-world problems 79, 215-16
transport economics 17
transportation costs
financing 140, 171
reducing 3
transportation costs/pollution costs trade-off
community recycling centre case study 215, 236-38
externalities 193, 199-201, 202, 204-5, 211-12
land-use planning 250
policy recommendations 255
transportation costs/production costs trade-off
economic geography 19-20, 105
facility location analysis 75-76, 93, 105, 127, 135
SPLP 75-76, 78-79, 92-93, 105, 106-7
transportation infrastructure
availability *versus* use 258
developing countries 76-77, 94
as ineffective policy instrument 104
network shape 39
policy, and network morphology 104-5
role of public capital 103-4
spatial and economic properties of 3
as spatial planning instrument 39
transportation networks *see* networks
transportation policy, history of 92
triangle inequality rule 146

UFLP (uncapacitated facility location problem) 6, 10, 23, 175, 176, 192
uniform delivered pricing policy (*U*) 176, 178-80, 182-87, 187-90
uniform mill pricing policy (*M*) 178-80, 182-87, 187-90

vertices *see* nodes
Voronoi diagrams 153, 158

zone pricing policy (Z) 177-78, 180-81, 182-87, 187-88